SPIRITUAL BEING HUMAN

THE SCIENCE OF SPIRITUALITY

KELLY PAARDEKOOPER

CONTENTS

FOREWORD

When I began my dive into spirituality, like most I know who go down this path, I was coming out of a dark night of the soul. The confusion that confronted me, knowing there was something more out there, something beyond religion but also so easily explained, was nowhere to be found. I went on a search for what could explain it in a way that was easy for this science failing human to understand.

I was met with scientific evidence that I simply got more confused by. Instead of being lit up and left wanting more, I retreated into my human existence for a little while.

It would have been a great pleasure to have stumbled across Spiritual Being Human at the beginning of my travels inward, deeper into myself, as this book just makes sense. Yet, meeting me now at this point in my journey, it is rewarding on a different level. It will not matter at what phase you are at, the knowledge will benefit you.

Whether it be your desire to understand how the science explains what we know to be a deeper truth or if it's just a better understanding of which "Claire" you might connect mostly to, or it what the whole "vibe" catchphrase is about, this book allows us to understand all this and more.

To think there are a plethora of books on the market about spirituality, with some so confusing they need a teacher's aide book to help explain them. Then there is this end of the scale where one can dive into any chapter and learn in layman's terms, or rather my terms – how it simply makes sense.

I am not lost in the science here, and yet, sometimes it goes deep into the science. Kelly just has a way with her knowledge and her background to explain so we can understand it. We can dive into the actual experience of being human in a spiritual way. Connecting to all parts of us and not limited to just talking about our subconscious and our heart brains.

Beyond the scientific learning is the lessons – about past lives, death, reincarnation, and the tools you should pop into your kit to help you along this path of human experience in a spiritual way.

It is encouraging to know that Kelly saw this gap in the market, and now I, too, can grasp some of the aspects of spirituality that were beyond my understanding until now.

-Brigid Holder

INTRODUCTION

For so long I've been a woman caught between two worlds, with one foot in the scientific world of evidence based medicine, and the other knowing there is something beyond that which we can perceive, a deep yearning to acknowledge and connect to my deepest self, an awareness below the surface that, until recently, I could not describe. For a long time, the only exposure I had to this deeper wisdom was in the form of things I saw on television. The closest thing to describing what I was experiencing was the word magic, to the point that I strongly identified with witches as they were portrayed on many TV programs. Going through life feeling like a witch in hiding, or perhaps a wizard like *Harry Potter*, awaiting a letter to tell you of your true birthright, isn't particularly socially acceptable so it wasn't something I talked about. The last thing I wanted was for my peers to think I lived in a fantasy-land full of witches and wizards, and maybe even fairies and goblins. Over time I suppressed this deep yearning in me and conformed to mainstream society.

Perhaps if there had been role models in the media showing that intuition is an innate layer of sense that we're all born with, or talk in schools of science being more than predictable Newtonian laws, I might've seen that I wasn't strange. This inner knowing in me wasn't just a result of daydreaming and watching too much *Bewitched*, *I Dream of Jeanie* and *Charmed*. Maybe I would have actively connected to my innate intuitive wisdom, and I would've been more open to looking within for answers rather than always reaching for books. But the love of research and facts and figures may not have come to be, and I wouldn't be here writing this book. Despite the sense of wasted years at not acknowledging and embracing my deeper truth, nothing ever happens by accident, everything has unfolded exactly as it is meant too.

So, what was the tipping point for me to sidestep off the treadmill of rational thinking, as per society's expectations? It was the beautiful mother energy that awoke within me. I can reflect on it and call it beautiful now, but the roller coaster of the early days of mothering (and even some of the more experienced days now) were anything but beautiful. Trying to adapt to life as a new mother, maintaining my career and attempting to keep my relationship together was all too much. My perfectly constructed world started to crumble. The best way to describe how I felt was that emotionally, it was like being stripped bare and standing naked in headlights. Being so raw and vulnerable saw me turn inwards, to that deeper self, to cope. And there, like tumbling down a rabbit hole, my world opened up in ways I could never imagine. The Universe laid out breadcrumbs for me to follow. Perhaps the breadcrumbs had already been there waiting for me to notice, but it was only after being so emotionally open that

I was able to see what was right in front of me. I found a course (or a course found me) that helped women connect with and grow their intuition whilst all supported by scientific evidence. Finally I had found my people. They understood me. And every experience I had was backed by my long time friend, science.

Like *Mulder* and *Scully*, I had been searching for the truth. Even when I was conforming to society and university and the black and white world around us, at some level I had known there was a deeper truth out there. In this course, I finally found what I'd been looking for, and the answer came in quantum physics. I actually studied (and did quite well at) physics in high school and touched on a fair bit of physics at university. But the quantum truth was totally new to me.

Since discovering this wonderful world of quantum physics, I have felt a deep calling to collate all the scientific evidence (not just quantum, but everything) out there into an easy to read, very accessible book for all those who feel a similar yearning to the mystical but are overruled by their reasoning, thinking, analytical mind. No longer should people feel caught between two worlds. The answers are out there. You are not "airy fairy" "hippy dippy" "woo woo"... (insert all the crazy words you can think of to negatively stereotype people who believe in this "stuff"). You are totally and completely sane and connected to reality. In fact you're probably more sane and connected than those who like to label us with these stereotypes.

To paraphrase the words of Dr Joe Dispenza, we are now in an era of demystifying the mystical. No longer is it ok to have people, gurus, whom we place on pedestals, showing us "magical" acts and asking us to have faith in them being real. Society is now in an era of having information at our fingertips, readily available at the touch of a button. We are now information and truth seekers. We have broken down the separation between us and the gurus. We are our own truth seekers and no longer accept everything in good faith.

Unfortunately the flip side of this is that we (society) have lost inherent faith and become more and more cynical. I know this because it is the path I walked myself.

Fortunately, my innate knowing has been ever present to guide me back to this path. As early as I can remember, I've always felt a connection to the mystical, I've known somehow that this is part of me, an awareness that I have a direct line to the Universal Consciousness.

Others may prefer to substitute phrases or terms such as The Divine, Spirit, Energy, Source, Infinite... Whichever word feels right for you, please substitute in places where I refer to Universal Consciousness.

In researching this book, I found that each chapter could easily be expanded into a multitude of books delving into each subject. It was a challenge to prevent myself from venturing too deeply into each topic. The aim of this book has always been to present the information and available evidence substantiating all elements of the mystical, in a straightforward, honest and easy to understand manner.

All the evidence is out there. Countless people have devoted their lives to studying these topics. But for every article/study in which I found substantiating evidence for a belief, I was able to find another squashing it. This is the interesting thing about science, it is designed by its very existence to prove things beyond reasonable doubt, yet thanks to the laws of quantum physics and the act of observation with intention, the outcome will always be skewed by our vantage point.

If you would like to delve even deeper into the science of spirituality, I have included multiple references at the end. But please note, I have tried to keep my referencing to a minimum throughout this book. I do not want you to feel like you are back at school, reading a heavily detailed textbook.

So please enjoy this book for what it is, a joyful exploration into the science of spirituality and an explanation, as I see it, of who and what we are as human beings.

1

THE SCIENCE CHAPTER

No doubt you've heard of the phenomenon of having a sixth sense. What if I told you that's absolutely rubbish. But hold up, this is a book to talk about the science of spirituality so how could I possibly be dismissing sixth senses? Well, it's not a full dismissal, more of a deeper explanation and a shift in understanding. Some call the sixth sense, ESP (extra sensory perception). Really, I think it should be called DSP, a deeper sensory perception, as it lies in our fundamental senses. But if we're going to continue to call it ESP, let us mix up what the acronym stands for and call it; Everyone's Super Power. Everyone has ESP. Another name for this is intuition, and we just haven't all been shown how to be open to receiving it. But to go into how this works, we first need to define intuition. Because what a lot of people are describing when they refer to ESP, is actually our intuition. Intuition lies in energy, in the waves of imperceptible "space" around us, within us, everywhere and in everything.

If we think of the world around us, all that we see, touch, and smell etc. This is the physical world, the reality in which we live, composed of finite molecules and atoms. It is these atoms and molecules that our senses detect, the physical stimuli that is matter.

Every piece of physical matter in this world can be broken down into its tiniest building blocks, and it is here, in these tiny atoms, where we find the foundation of intuition.

If we look specifically at the human body, we are made up of a multitude of different organs. These organs are made up of cells, which are made of organelles. The building blocks of organelles are molecules, which in turn, are made of sequences of atoms. Now this is where it gets interesting, at the atomic level… atoms are proposed to be made of over 99% "empty" space. The journey this book takes us on, has always been intended to break down the truth as I see it, rather than just tell you stuff and expect you to believe and accept it on face value. So, how can an atom be 99% empty space? And what are the implications for us as human beings, if our fundamental building blocks are actually only 1% physical matter?

If we look at the structure of an atom, it is made up of subatomic particles called protons, neutrons and electrons. Atoms are measured in a unit of length called an angstrom. Angstroms are 1 10 billionth of a metre, which can be expressed as $1x10^{m}$ (expressed as a fraction that means 1 over 10 with ten zeros on the end!). So infinitesimally tiny. Now, neutrons and protons are 1/100 000th times smaller again. The interesting thing is that atoms aren't then filled with hundreds of thousands of these subatomic particles. Depending on which atomic element we are looking at, there are only, at most, a hundred or so protons and neutrons combined. And as for electrons, they don't even factor into the equation as they are 1/10 000 times smaller again than protons and neutrons.

Have I lost you yet? Stay with me, it's worth trying to wrap your head around this science as it really clearly helps demonstrate just how much "empty" space there is, and of course, how it relates to intuition.

If you ever have reason to look at a periodic table, the atomic number of an atom represents the number of protons. If you then look at the atomic mass (usually expressed as a number below the atom name) it is the number of protons plus neutrons. Electrons aren't factored into this as they are so inconsequentially small by comparison. But for the record, and to round out the Chemistry 101 lesson, for an atom to be stable, it needs to have the same number of electrons as protons.

So, that means, these subatomic particles only make up 1/100 000th -1/1000th of an atom, the rest is seemingly empty. Really the specific numbers don't actually matter, I explained them only to give you an idea of just how immensely vast the "empty" space is of an atom.

So, everything in this universe is made up of these atoms, These minute little building blocks that are primarily empty space. By description then, that makes us, as human beings, predominantly empty space. Now I'm sure you know plenty of people who do appear to be nothing much more than hot air, but that's not what I'm meaning.

Have I got your mind going in circles yet? Let's look at the empty space. Scientists have been able to show that it's not actually empty. This is where quantum physics comes into play.

Things that are macroscopic, i.e. meaning that we can see them with the naked eye, behave according to Newtonian Physics. This means that they act in a predictable manner. This branch of physics is called Newtonian after Sir Isaac Newton, a physicist from the 1600-1700s. He is considered one of the most influential scientists ever as he first broke down visible white light into the coloured light spectrum as well as having proposed the The Laws of Motion that explain the concept of gravity. It's these Three Laws of Motion that are the basis of the Newtonian approach to physics. Big stuff follows Newton's Laws or classical mechanics. Tiny stuff, the things we can't see even with a standard microscope, don't follow these rules. They follow what's called quantum physics.

QUANTUM PHYSICS

Here's where the fun really begins. The foundation of quantum physics is that subatomic particles (the protons, neutrons and electrons I introduced you to before) actually behave as finite individual particles, as well as like a continuous wave. You may have heard this described as wave-particle theory or duality. Multiple extremely well known scientists have contributed to this field of research, eg. Max Planck, Albert Einstein, Neils Bohr. Unless you're a science geek like me, these names may not mean a lot to you, but let's just say they're some of the A-lister, red carpet type celebrities of the science world. There have also been a significant other list of prominent physicists to have had a role in quantum theory as we know it now. The fact that I don't mention them is not to say that their work isn't important, just that the names may not have relevance for you, and I think I've name dropped enough to give you an idea of how important the discovery of wave-particle duality has been for modern science as we know it.

My first lightbulb of understanding into how this relates to modern spirituality came in the form of learning about The Double Slit Experiment. It is one of the most famous and consequential experiments in physics and was first conducted in 1801 by Thomas Young. This experiment was first performed with light particles, as solid beams of light. Despite the individual particles being too small to observe with the naked eye, it was possible to see the impact of them as a whole. More recently, the study has been replicated with electrons by Davisson and Germer in 1927. Since then it has been replicated countless times so there's no denying the science here.

The premise of the experiment is that when macroscopic particles (large, visible matter) are fired through a barrier with two, parallel slits (small gaps in the barrier), the particles land on the wall behind in a predictable pattern matching that of the location of the slits, i.e. in two parallel lines. When the experiment is then conducted with microscopic particles (e.g. electrons of energy) the particles act in a completely different manner. It doesn't follow the predictable path of macroscopic, physical matter. Instead, energy follows every possible path, simultaneously, all at once, including that of formlessness (pretty crazy to try to imagine right?) and presents as a wave of motion. This in itself is the basis of quantum mechanics. Macroscopic matter follows the predictable, if you will, black and white world of Newtonian physics. Microscopic particles (which remember, are the building blocks of macroscopic particles) follow the limitless potential of quantum physics.

Now this is where it gets fascinating, when the scientists went about trying to observe the exact path each microscopic particle followed to see how this was actually possible, the wave function collapsed and the particles reverted to Newtonian motion (so they reverted to hitting the screen behind in two parallel lines in accordance with the position of the slits). Scientists have performed multiple variations of where the observations took place and even in which time frame they occurred. The outcome of these variations showed that ultimately, the particles always behaved in accordance with the belief system of the observer. What we can deduce from that is that energy is energy, until it is observed, after which it becomes matter.

The exciting consequences of this are multiple, actually infinite and limitless. It highlights that we are energy first, matter second. Everything is always energy until we apply focus to it, then it becomes physical matter. And remember, this isn't a recent discovery, Young first stumbled upon this over 200 years ago, and it was still nearly 100 years ago that Davisson and Germer replicated and expanded on the initial discovery. So why isn't this part of our mainstream understanding of the world in which we live...?

Somehow through my high school physics lessons and what we covered at university, I missed the fundamental cornerstone of quantum physics. Perhaps if I'd connected the dots then, I wouldn't have taken so long to welcome and embrace my own spiritual, mystical side. Discovering the full scope of quantum physics in my 30s was instrumental in me no longer feeling as though I had a foot in each world, instead it allowed me to see the Newtonian world in which I physically exist is very much the exact same as the one in which I feel, see and know energy. The world in which my intuition communicates through all my senses and my heart is connected to everyone and everything. There is no separation, they are one and the same.

Well done if you've made it this far, that last section of science started to get pretty intense and most likely quite confusing. The ultimate takeaway is that more than 99% of everything (including human beings) is seemingly empty space that constantly has energy occupying it. It is this energy that is always interacting with the world around us. We don't need to see it to know that it's there, but through the laws of observation, when we look for it, we will perceive it. This is how our intuition works.

Our primary senses quite clearly detect macroscopic particles, but they also detect microscopic and even subatomic particles but this is much more subtle than the macroscopic. We can perceive energy. The simplest way to demonstrate that is to think back to a situation where you knew someone was looking at you, or you may have "felt" someone walk into a room even though you had your back turned to them. Your rational, analytical brain probably dismissed at the time with some Newtonian justification like "they must've made some noise that I heard". But what if there wasn't a perceivable trigger to alert you? It is possible that you could tell they were there by a deeper awareness in your senses.

Each of our five primary senses have an intuitive layer called a clair (after the french word for clear).

Clairvoyance - clear sight
Clairaudience - clear hearing
Clairgustance - clear taste
Clairnascence - clear smelling

Clairsentience - clear feeling

Now I know I said earlier that there's no such thing as a sixth sense. Well I was being flippant, there is actually a sixth sense, but like the other five just mentioned, it's not something special, merely a deeper layer to an awareness that we already know exists. The sixth is

Claircognizance - clear knowing

If you have trouble remembering them, think of these little word associations…

Clairvoyance - voyeurs can see
Clairaudience - is it audible?
Clairgustance - taste is connected to your gut
Clairnascence - "Nascence" "Nose
Clairsentience - can you feel a sensation?
Claircognizance - cognition in our brain, our place of knowing

Everyone experiences and receives intuitive information slightly differently. I am extremely fortunate to be surrounded by a great many people who are deeply in touch with their intuition so I asked them to describe in their own words what it's like for them to receive information through their "clairs".

(clairvoyance) "I see stuff (in movie type format in my minds eye) - Think of it like an old school projector on a screen, only I am the projector"
(clairvoyance) "...seeing - through the darkness behind my eyes"
(claircognizant) "information that exists in my brain

and seems like its always been there, like I learn it at school"
(clairsentience) "goosebumps AKA truth bumps"
(claircognizance) "I know stuff that just pops into my head"
(clairsentience) "I can feel in my body what other people are feeling in theirs"

Many people will find that once they tune in and pay attention to these senses, they'll discover they've already been receiving intuitive hits via these senses, but perhaps didn't give it enough credit. We are all born intuitive, it is innately part of who we are. As we mature and age, our focus becomes less on what's already inside, and more on the external world around us.

Initially you will find that you have one, or perhaps two, dominant clairs. When you open yourself to your intuition, information will come through primarily via this sense. As time goes on, each of your other intuitive senses will grow stronger and start to become more noticeable.

Before I even realised I was intuitive, I would have an uncanny ability to know stuff that other people didn't. I mostly thought that it was because I was a good student and I must've learnt it at school. My rational, conscious mind reasoned that everyone else must've just not been listening that day. Now that I look back on it, I realise how crazy that is. How on Earth could I have been the most attentive student ever? And what are the chances that no-one else ever learnt these same things? As you may have guessed, my dominant clair is claircognizance. Over time, I started to feel things (clairsentience), sometimes as goosebumps (truth bumps) when I would hear something or experience a thought that was significant for me to pay attention to. I also began to notice that I would be able to feel sensations or tingles in my body that correspond to something going on for a person I was with. Over time, the more I have embraced and used my intuition, the stronger it has become. I'm no longer bound by time or space to be able to tune in to others. I personally know of an instance where someone, who is clairsentient, suddenly had their finger bent at an abnormal angle. They later discovered that this had occurred at exactly the same time a loved one broke their finger (and this is a first hand account from someone I trust implicitly, as is the witness who was there to substantiate the claim).

Please know if you have any reason to doubt or question your intuition, this layer of awareness is real, your intuitive senses have names and it is ok to start to use them to acknowledge and describe your own intuition, more freely. You are not imagining it. Your physical senses detect things on a Newtonian level, as well as on a quantum level. The great thing about intuition and your "clairs" is that the more you use, and pay attention to them, the stronger they will grow.

2

IT'S THE LAW

So you might have heard about the Immutable Laws... maybe, maybe not?

If the term doesn't ring a bell, I'm quite sure that these principles will still be familiar to you. The Immutable Laws of the Universe are fundamental principles that many take for granted without understanding the basis of them. They are the order and structure that keep the Universe in perfect balance and harmony.

So let me take you through the seven Immutable Laws and give you a little insight into how and why they are so important.

The Immutable Laws are sometimes also referred to as the Hermetic Laws or the principles of the Universe. They first came to light in the modern era in a book called The Kybalion. First published in 1908, The Kybalion is reportedly based on the teachings of Hermes Trismegistus, a legendary ancient Greek author and philosopher (are you starting to get the idea that nothing is ever a new idea, just newly presented in different languages or targeted for varied audiences). The first mentions of these laws have been found in Corpus Hermeticum (reportedly written by Hermes Trismegistus) as well as parts in Nag Hammadi library from 4th century Egypt.

This overlap of appearance of these laws in both ancient Greek and Egyptian cultures suggests either an even older fundamental teaching (it's worth noting here that Hermes Trismegistus is thought to be a combination of the greek god Hermes and the Egyptian god Thoth), or perhaps a deeper knowledge of the absolute truth to which we as society are still waking up. Either way, these laws that were true thousands of years ago, still hold true today.

Now technically there are seven Hermetic Laws, three of which are immutable and four are mutable. Understanding this differentiation was pivotal to me. When I first learnt about the Hermetic Laws I could see, feel and know the truth in the first three, yet the last four laws seemed to contradict the first three. The Immutable Laws are fixed and absolute. No matter how the universe changes and evolves, these three laws will not. The Immutable Laws exist on all levels, physical, mental and spiritual. The four mutable laws however can be transcended and eclipsed by our own internal work and understanding of the first three laws. That is not to say that the final four laws should be ignored. I prefer to think of these as the introductory laws as they are the laws of the universe we tangibly see around us in our 3D reality. It is only as we start to do the work and truly understand that we are spiritual beings, existing in 5D, that the first three, the Immutable Laws become our governing truth.

The seven laws are the laws of mentalism, correspondence, vibration, polarity, rhythm, cause and effect, and gender.

1. THE PRINCIPLE OF MENTALISM

"The All is Mind; the Universe is Mental."

This law tells us that the mind (consciousness) is the foundation of the Universe. Nothing else is real, it is all a creation of the mind. Our reality is a projection of our consciousness. And this consciousness is universal, it is all of the one mind.

If we go back to the old saying of "what came first, the chicken or the egg?" The truth is neither, it is consciousness. Everything else, including the chicken and egg are created from consciousness.

What this law also tells us is that this one mind connects us. We are part of this one mind. The only thing that truly separates us is the power of our own free will. Our free will influences the projection of our own individual realities.

2. THE PRINCIPLE OF CORRESPONDENCE

"As above, so below; as below, so above."

"As within, so without; as without, so within."

This law is the harmony between everything and highlights the lack of separation. Everything around us is replicated. It also shows the underlying interconnection between everything and everyone. What we see, know and understand helps us see, know and understand everything else.

Perhaps the easiest way to understand this law is to look at the spiral structures that appear in nature, such as the spiral of a shell or the growth pattern of petals on a flower. Now without getting too detailed, they follow a precise mathematical formula (for my fellow geeks, it's the Fibonacci spiral) that has been found in everything, from the tiniest snail to the largest galaxy; as above, so below.

We are all divine parts of this universe. Applying the principle of correspondence one step further tells us that not only are we part of this universe, the universe is part of us. Whatever your faith and higher belief, knowing this principle reminds us that there is no separation between us and all divine higher power.

3. THE PRINCIPLE OF VIBRATION

"Nothing rests; everything moves; everything vibrates."

Vibration is a word that gets used quite a lot in spiritual circles and this principle brings it back to basics. We talk so often about vibration because everything is a vibration.

Vibration is such a huge topic and so important that I've given it its own chapter "It's a vibe".

4. THE PRINCIPLE OF POLARITY

"Everything is dual; everything has poles; everything has its pair of opposites; like and unlike are the same; opposites are identical in nature, but different in degree; extremes meet; all truths are but half-truths; all paradoxes may be reconciled."

This is the first of the mutable laws, the laws that can be transcended by our own spiritual growth.

This law tells us that everything has an opposite and is based on duality. Deeper than this, the principle of polarity reminds us that through opposing forces, we attain balance. And opposite forces are part of the same. For example, cold and hot are part of the same, cold simply is the absence of heat.

Polarity also gives us a reference point for understanding. Without bad, there would be no good. Polarity is an important principle so long as we remember that opposites are still part of the same.

5. THE PRINCIPLE OF RHYTHM

"Everything flows, out and in; everything has its tides; all things rise and fall; the pendulum-swing manifests in everything; the measure of the swing to the right is the measure of the swing to the left; rhythm compensates."

This law follows on from the principle of polarity in that between all opposing forces, there is an underlying connection of rhythm. What this means for us as human beings is that everything is connected, and that connection is fluid meaning that there is natural change and movement. It is natural for us to want to resist change but this principle highlights that change is an inevitable part of life. Just as the moon cycles each month, the seasons each three months and the planet each year, so to do we, and the lives around us, change.

Embracing this principle of rhythm and change affords us comfort during the difficult times as we know they won't last forever, and deep gratitude for the good times as we know that they too may pass.

6. THE PRINCIPLE OF CAUSE AND EFFECT

"Every cause has its effect; every effect has its cause; everything happens according to law; chance is but a name for law not recognised; there are many planes of causation, but nothing escapes the law."

There are no accidents. Everything happens for a reason. This law tells us that every single thing that happens has an underlying cause or will instigate an important effect, everything is part of a larger picture of cause and effect. The challenging thing here is that quite often we, as human beings, can't see the bigger picture. If we liken life to a jigsaw puzzle, it's rare for us to be able to see much more than just a few assembled pieces. But please trust that everything has a reason, it is all part of the grand design of your human experience.

7. THE PRINCIPLE OF GENDER

"Gender is in everything; everything has its masculine and feminine principles; gender manifests on all planes."

According to this rule, everything has a gender, everything is binary, either masculine or feminine. This law is thought to apply to absolutely everything across all planes.

I'm going to be a little controversial here, after all this book is about breaking down fact and fiction and calling things out for what they are. This law is starting to show signs of no longer being valid. We often hear statements like "society is awakening" or "evolving" and this law may be evidence to the fact. Society is starting to no longer see the world as separate genders. Although, in saying that, we still have an incredibly long way to go.

In 1908, I can see how the world would've been perceived to exist in gender, I mean even latin based languages assign gender to everything e.g. in Spanish, words that are "feminine" end in -a whereas words that are masculine end in -o. Somehow a chair *"silla"* is female yet a book *"libro"* is male. Human beings aren't that black and white. We are not always binary. Many people do not identify with the gender in which they were born, and others do not identify with gender in any form. This law is no longer valid. We are beginning to evolve beyond this simple binary allocation of gender. Perhaps this law is an example.

So now that you know the Hermetic Laws of the universe, what do they actually mean? And how do you implement them into your day to day life? The way I see it, the four mutable laws (polarity, rhythm, cause and effect and gender) are quite tangible and ingrained in society already. You don't need to do anything further to understand or embrace these laws other than knowing they exist.

Gaining a deep understanding of the three Immutable Laws (mentalism, correspondence, vibration) however, will free you from the confines and limitations of your perceived reality. To know that all is of the mind allows us the freedom to know that we create our own reality. Understanding that everything comes from the mind affords us the freedom to choose how we create our own reality. We gain insight and understanding about the inner workings of the smallest possible things as well as the grandest concepts beyond our reach thanks to the law of correspondence and we have the tools to understand how to change our reality thanks to the law of vibration. Everything is but a vibration. By being aware of our vibration and tapping into the one mind of Universal

Consciousness, we can access the limitless truth that we are all spiritual beings having a human experience.

3

SPIRITUAL BEINGS HAVING A HUMAN EXPERIENCE

The paradigm of being a spiritual, energetic presence going through life in a physical body with all its lessons and limitations, to me is the very crux of the meaning of life. And we are all experiencing this life, together.

Funnily enough, if you look inside the word together, you can see the word ego, backwards. The biggest barrier to truly working together in anything is the natural instinct to push our egoic agenda to the forefront. We all need to reverse our egos and come together, for together we shall rise.

So what exactly is ego? I was first introduced to the concept of ego in my first year of psychology at university. According to Sigmeund Freud, the ego is the go-between, the mediator between the id (the part of us that is needed to ensure our primal needs are met) and superego (the moralistic aspect of our personality that develops as we mature). The ego acts to strike a balance between these two, often opposing forces. If we predominantly acted in the id, we would be purely focused on our fundamental needs with total disregard for the needs of others. We would also be oblivious to what is socially and morally acceptable. To act purely from the superego, on the other hand, would have us more often than not, neglecting and sacrificing our own needs. The superego would have us so focused on everyone else and doing the right thing that we would never be able to succeed as individuals. Freud described the ego as our guiding force so that we may act in balance and harmony in this world.

In popular culture however, ego has come to mean something very different. Thinking back to the approach and terminology of Frued, society tends to use the term ego as a way of describing the id. If a person is egoic or being dominated by their ego, they are singling themselves out from others and pushing their own agenda. The opposite of this, is to be acting in the super-ego, to be operating for the good of others or society/humanity as a whole.

Same, same but different. It's all the same concept but different words are used.

Let's add another layer of understanding... We are all spiritual beings having a human experience. The ego separates us from our spirituality, from our spiritual self. When we no longer allow ourselves to be dictated by ego, we open to our deeper selves, where we find that everyone and everything is connected. We are in fact, all one, there is no separation. There is only ever one of us here.

HOLOGRAPHIC NATURE OF THE UNIVERSE

I have been on the journey of exploring and awakening my own spirituality for some time. Over the years, I have heard numerous teachers refer to the holographic nature of the universe, to the point that it's felt like a little catch phrase to explain away certain things. So I delved deep, deep into the science to try to understand both how and why this concept is relevant.

Now although I am a self-professed science geek, I am not a theoretical physicist so the detailed explanation of the holographic nature of the universe is still challenging to totally get my head around. However, I understand enough to be able to give you an idea of the principles behind it and the resulting significance for how we perceive our human existence in this world.

Firstly, let's define exactly what a hologram is (because I always just thought it was cool sci-fi technology that allowed Princess Leia to be able to get a message out of the Death Star and in front of Luke Skywalker). A hologram is a two dimensional image that appears to have a third dimension, depth, when it is viewed. So what that means is that it a 2D image that actually looks as though it is 3D.

Over the years there have been a number of brilliant minds who have progressively built on each other's work to get to where we are today and what is known about the holographic nature of the universe. Ultimately, it all started with black holes. I'm sure you've heard of them, areas in space where the gravity is so immensely dense, anything nearby gets pulled in and nothing escapes, including light. A key discovery of black holes is that what goes in is exactly proportional to the surface area of the black hole (as opposed to the volume inside, which is what would be expected). Stephen Hawking added to this discovery when he found that information that went into the black hole was destroyed and then leaked out as Hawking Radiation.

If you think back to one of the fundamental principles of energy you may have learnt in high school science, nothing ever disappears, it merely changes form, this makes Hawking's discovery a paradox as information cannot be destroyed.

Gerard T Hooft came along in the latter part of last century and built on this further to discover that what enters a black hole will also be energetically imprinted on the surface. His theory describes that what enters the black hole exists inside in its three dimensional (3D) form. At the same time, identical information exists on the surface in a two dimensional (2D) form. This is where the fancy explanations start to kick in. This is described as holographic space time, when a 2D surface encodes the properties of a 3D interior. This theory proposes that not only can we describe the properties within by looking at part of the surface, but we can actually look at any part of the surface to describe the interior. All the information is encoded everywhere. There are three mind-boggling things to note here, a) everything is everywhere, b) this is an example of two things existing in different places at the exact same time, and c) 2D information allows for an extra, third dimension.

The ability for something 2D to appear 3D is explained by string theory. This is yet another branch of quantum physics that may leave you feeling like your brain itself is a jumble of strings so I'll try to keep it brief and dive into only the stuff that is relevant for our greater understanding. String theory proposes that nothing is a definable separate entity, rather points connected as though by a piece of string. The lengths of theoretical strings can vary without changing the detail of the information, meaning that the rules of interaction in the environment are unchanged, merely scaled. So you can change the size/scale of the image without changing the contents (a bit like enlarging a picture). The way strings scale in size allows for the addition of an extra dimension, the further they are apart, and the greater the scale of the image (the more the image has been enlarged), the more chance there is of another dimension appearing in the spaces between the strings. Keeping in mind that in terms of physics, a dimension is simply a direction you can move in, so when we add an extra dimension, this refers to adding another direction in which to move. It was found that when an extra dimension emerges, so too did gravitational pull.

Gravity emerges in a holographic way. The reality is two dimensional but when we view it, a third dimension seems to appear, just as per the definition of a hologram, a 2D image appearing in 3D.

So how does this relate to spirituality? Spiritual teachers have extrapolated this to propose that nothing around us is actually real, rather a projection of reality. If you've ever seen the movie, The Matrix, with Keanu Reeves, it's a little like that. We exist in a projection, an illusion of reality. The world around us is one that we shape with our own observation and expectation.

If we accept this premise as fundamental truth, then everything we do and everything we are is of our own making.

Additionally, accepting the rules of string theory, and the quantum theory of energy, discussed previously, the empty space that separates us from everything else, isn't actually empty space. It's abundantly full of energy that is constantly in motion. This energy connects us all so that we are not finite, separate beings. We are all connected, all the time. Everything we create in our reality, we do likewise for others. As they do for us. Not only is it our perception that creates our reality, but also our connection to everything around us.

NEAR DEATH EXPERIENCE

One particularly interesting element of being human is the finite span of our life, and the recollections of people who have momentarily left their human existence, experienced themselves fully in the spiritual realm and then returned to their human body to tell the tale.

Specifically, I'm referring to the reports of people who have physically died and been revived, or had what we call near death experiences (NDE). Through these reports we can start to join a few more dots about this holographic nature of the universe and our perception of reality.

Until recently, I'd always had the belief that it's quite a rare phenomenon, but the more I started to talk to people about it, and ask questions, the more I discovered that NDEs are surprisingly common.

Like many things, I grew up accepting that NDEs are real as a relative had experienced one in his early-mid 20s. Whilst I knew of his, I thought it was so rare that the only way I'd ever hear of anyone else having had an NDE was through the media. I truly thought that only a handful of people have had this experience.

Recently however, I've discovered so many first hand friends and acquaintances who have experienced something similar. I even realised that I had perhaps had my own NDE whilst giving birth to my daughter. Until recently, I had blocked out aspects of the final minutes of labour in which I felt I was on the cusp of dying. I could feel myself hovering in and out between the birthing suite and a place of calm, peace and expansiveness. It was only my husband telling me that our baby had arrived safely that pulled me completely back into this reality. Perhaps NDEs aren't nearly as uncommon as I had thought.

One of the highest profile NDEs is that of Anita Moorjani which she shares in her book "*Dying To Be Me*". Anita's body was dying of lymphoma, a form of lymph node cancer, and reached a point where she was in hospital as her body began to shut down. Doctors had given her only hours left to live. Loved ones were preparing to say their final goodbyes and her brother was on a plane, desperately trying to get to her in time. Anita reports that in that time, her body drifted in and out of consciousness, hovering between two worlds. She describes the other world as having crossed over to another dimension where she was engulfed by an overwhelming sensation of love.

Ultimately, Anita was given the choice to come back, or not. She chose to come back when she realised that heaven is a state of being rather than a place. She was shown that there is no separation, she was able to see everyone and everything at once. Time and distance failed to exist, she was part of consciousness. Anita gives the analogy that it was like being in a the most expansive and vast, but incredibly dark warehouse with only a flashlight to be able to see directly in front of her. Everything possible existed in the warehouse, but it could only be seen if the flashlight was shone in its direction. Her NDE and experience of entering another dimension was like having all the lights turned on in the warehouse so that she could see everything clearly. Ever since the experience, Anita reports that her perception and awareness has significantly changed. Things that couldn't have imagined knowing seem to already be in her awareness. She even recounts conversations that happened in other rooms of the hospital whilst she was drifting in and out of consciousness.

A key part of her story is that during her time in the alternate dimension (as she called it), she was shown that if she made the choice to return to her physical body, her body would begin to heal. What happened was actually truly remarkable, she went from multiple organ failure and the process of her body completely shutting down, to having an 80% reduction in lymph node size within four days. Within two weeks, biopsies were unable to find any evidence of cancer. That is absolutely mind blowing!

Through all my discussions and interviews of others who've experienced NDEs, no one else has had such a clear and profound experience, most report seeing light and a sense of nothingness, but not empty, rather full and complete but without physical "stuff" or stimulus.

Many people report having an awareness that they have to return to their physical body in order to do more work here on Earth. Perhaps this is why Anita's experience is so profound, because part of her work is to share her story and to teach us about our own human experience.

BEING HUMAN

There are so many amazing, wonderful lessons and truths to be found as spiritual seekers. In this experience of being human, it is about learning the lessons for ourselves. I know I've certainly read an absolute multitude of books and keep waiting to find that magic formula to unlock so many of the truths about which I’ve read. Yet, I still keep coming back to the fact that I am a human being. I have to remind myself that I am instead, a spiritual being having a human experience, and it's okay to occasionally get lost in that humanness. No one expects us to get it perfect, the only expectations are the ones we place on ourselves. The pressure of perfection to which we subject ourselves can be brutal, and at times harmful. I am giving you permission to embrace your humanness, to make the mistakes you need to make to allow yourself to take two steps forward, then one step back. That one step back is just as important as those two steps forward.

Wouldn't it be amazing to have all the answers and to no longer make mistakes in life. However, when we get to that point in life, it means there are no more lessons to learn.

At the risk of becoming a professional student in the School of Life. We don't know how much there is to learn until we start learning.

Living life as a spiritual being having a human experience is all about the journey, and not the destination.

4

THE HEART KNOWS

In 1638, William Harvey first proposed that the role of the heart is to pump blood around the body via a network of arteries. Can you believe that he faced ridicule and significant criticism for this until about 20 years later when his work was finally widely accepted?

Fast forward to the current day and we now know so much more about the heart, and yet we've only just scratched the surface as there is still so much more to learn.

Physically and anatomically, we know that the heart is the most unique part of our physical body. It is a muscle, an organ and an endocrine gland. The types of cells in the heart are typical of a muscle, yet by definition the heart falls into the category of an organ because it contains groups of tissues that work together to form a specific function, i.e. to pump oxygenated blood around the body. However, it also falls under the category of an endocrine gland because it has been found to secrete hormones. No other part of the human body is capable of performing all three of these functions.

In addition to its physical uniqueness, the heart is also considered to be the source of our emotion, wisdom and courage. How often have you experienced an overwhelming emotion and physically felt it in the centre of your chest? Perhaps the sensation of warmth and expansion that comes with feelings of love, joy and gratitude? Or have you experienced the awareness of pain and constriction in response to emotional hurt and betrayal?

The HeartMath Institute are pioneers in studying both the anatomy and the functionality of emotions in the heart. Specifically they explore what it is about the heart that allows it to have connections such as neural pathways to connect it to the brain to link our hearts to our emotions.

One of their key observations is that the heart acts independently as though it has its own brain. Traditionally, it has been thought to be acting in response to signals and messages sent from the brain. We all grew up being taught that the brain is the control centre of the body, it is in charge, and every other part of the body follows along.

Traditionally, it was thought that the brain is the single power centre of the body, sending instructions to all other parts. This is only a portion of the story. The heart sends just as many messages back to the brain in return.

HeartMath have also found that negative and emotionally draining experiences such as stress and frustration, negatively impact certain functions in the brain as well as causing changes in heart rhythm at exactly the same time. They noted that these changes in both the brain and the heart went on to negatively influence all other body functions. This suggests that the heart is not instructed by the brain but instead has its own ability to process information and influence the body.

When we experience different moods, our heart rate changes, but instead of the mood in our brain dictating how our heart functions, it's been found that the heart rate variation comes first. The heart rate varies, signalling our autonomic nervous system which then in turn causes changes in the brain. The heart dictates our cognitive state.

HeartMath have found that the heart is a "highly complex information-processing centre". It has its own brain, dubbed the heart-brain, capable of two way communication with our traditional brain, the cranial-brain. The heart is the control centre of our creativity, emotional processing, attention level and intuition.

There are reports of heart transplant recipients taking on the personality traits, emotions and likes and dislikes of the donor. This means that the transplanted heart influenced aspects of the recipient based on information previously stored within the heart. This suggests that it also has cellular memory.

Neurocardiologists (a combination of cardiology and neurophysiology) have discovered that the heart is made of a network of nerve signalling structures (ganglia, neurotransmitters and proteins) that are very similar to that in the head-brain. This neural circuitry in the heart-brain allows it to be able to learn, remember, make decisions and sense and feel independently from the head-brain.

HEART RATE VARIABILITY

If you've ever had your blood pressure taken, you'll know that heart rate is measured as a number of beats per minute. But did you know that during that minute of measurement, there is variation and fluctuation in the rate between each beat. The heart rate is an estimation of the average of the beats throughout that time, yet heart rate variability (HRV) is a measure of how much variation there is between each beat. It is thought that HRV is an accurate measure of health and how fast we age. HRV is a reflection of the resilience of the heart to respond and react when needed. Measurements are highest in young children and decline as we age. We can think of the heart as suffering wear and tear as we age. This is especially so for stress-induced issues which have been noted to cause advanced decline in HRV irrespective of age. Abnormally low values have been associated with increased risk of a number of diseases and disorders.

To help combat this, HeartMath has discovered that we can consciously influence our heart rate and stabilise our HRV. This has immediate benefits to both our mood and our overall physiology (remembering that the heart-brain sends important messages throughout the body) as well as the long term gains in improving HRV. HeartMath have developed a specific called Heart Coherence, to achieve this.

HEART COHERENCE

It's worth noting here that Heart Coherence is not a form of meditation or relaxation. Heart Coherence is a specific process of cultivating feelings of love, joy, gratitude and appreciation, feeling those feelings anchor in your heart, then allowing them to swell and grow. This practice creates stable, smooth, balanced heart rate rhythms as opposed to the erratic and fluctuating rhythm of stressful emotions such as frustration and anger.

Step 1; Bring your attention to your heart. Notice the place your anatomical heart sits inside your chest and imagine your breath slowly flowing in and out of here. Find a calm rhythm of breath that is comfortably a little slower and deeper than your regular breath.

Step 2; Focus on positive emotions such as love, joy and gratitude. Feelings such as those that come from tasting a perfect cup of coffee or the feeling of getting into a bed made with freshly washed sheets are best. Feelings that involve others can sometimes be a little challenging for this practice as human relationships often have layers of feelings and emotions that can distract from the purity of love.

This practice only takes about 60 seconds to receive benefits. However, the longer you can cultivate these feelings, the better.

According to HeartMath, not only will this practice achieve balance within your body, it will also energise your system, access your heart's intelligence, and increase your intuition.

Yes, you read that right, our heart is connected to our intuition. By holding positive emotions, which then in turn stabilises our heart rate, we can increase our intuition? Yes seriously. This is something I have not only researched and studied, but also practiced, a lot!

Another way to think of the heart is that it's the "feeling brain", as opposed to the actual brain in our head which is the thinking brain.

Another fascinating trait of the heart is that it is the strongest source of electromagnetic energy in the body. HeartMath have found that it has an electrical field that is approximately 60 times greater than the brain, and the electromagnetic field produced is over 100 times larger that the field created by the brain. This electromagnetic field can be detected as far as one metre beyond the body, and that is only what the current measuring devices have been able to detect. It is predicted that as technology advances, we will discover that the electromagnetic field reaches far beyond this measure.

An electromagnetic field is a form of energy. All energy interacts and influences each other. So if you imagine standing one metre (or more if you're in a pandemic and "social distancing"), the electromagnetic field of your heart is going to come into contact with the electromagnetic field of the person beside you. Your energy will be mixing with theirs. Your energy impacts others, and vice versa. By consciously increasing our energy and our vibration, we can positively impact that of those around us.

Conversely we can also be equally impacted by the vibrational energy of others. This is an important thing to remember when we choose who we associate with. Unfortunately, in our current society, we don't always have the choice to be completely selective of those who surround us. Some people who choose to go by the label of empath (someone who empathically feels the feelings of others around us) may be very aware of feeling vulnerable to the energy of others. We don't need to be vulnerable. The electromagnetic field of our heart can act like a shield. The stronger the vibration of our energy, the more it will elevate and positively influence any lower vibrations in our environment. Along with heart coherence, holding the feeling, the energy of love in your heart is the most powerful way to increase your vibration and your heart's electromagnetic field. We will talk more about raising your vibration in future chapters.

LOVE - THE LANGUAGE OF THE HEART

At school there was this phrase "to be up yourself" or to be "full of it". I don't know if they were common phrases tossed around at your school, but they certainly were at mine. These were derogatory terms for loving yourself. This was in an era when we were encouraged to be self critical, to cultivate hatred for yourself because the last thing you wanted to be was "up yourself". Now that I'm an adult, I firstly find that phrase ridiculously foul and bordering on laughable, but the actual implications of the concept are tragic!

If you thought that you were good at something or that you were proud of, or if you felt good about yourself, it would invite criticism and derision from your peers.

I don't recall where I first started to believe this concept but it certainly feels like this has become a constant internal dialogue my entire life. During my childhood, I was doing everything possible to avoid being "full of it" i.e. I was constantly sabotaging my self-confidence. Fast forward to now and I have learnt that that is exactly what I need to be "full of it", full of self-confidence and most importantly, self love. Not in a boastful, bragging sense, but in believing in myself, being proud of myself, and above all else, loving myself unconditionally.

The absolute crux of getting the most out of this human experience is love. Love is the most powerful thing. It is more than an emotion, it is a state of being. You can't truly open yourself to loving anything else unless you have love within you. All other forms come with conditions and needs. When we already have this unconditional love within us, all other love is a bonus and does not need conditions or expectations. Imagine loving your family or partner without conditions. Envision totally and completely loving another human being without any expectation or need for that love to be returned. How liberating would this be? It is possible when we enter a state of unconditional love for ourselves. When the truly expansive state of this feeling becomes our natural state of being.

We see the world through the lens of our human physical form, and the centre of this physical form is our anatomical and energetic heart. This is where the love radiates from, the centre of our being. Even in the chakra theory of energy, it is no accident that the heart chakra is the very centre of the physical body. The heart is thought to be the axis through which all energy enters and leaves our body.

At our centre, we are love. If we don't cultivate and nourish this, if we don't embrace it as our natural state, how can we possibly truly allow that love to radiate out from the centre of our being to love another? Every-time I think of love as our natural state, one of The Beatles songs pops into my head..

> *"All you need is love, love, love is all you need…"*
>
> The Beatles

5

I'VE BEEN HERE BEFORE

REINCARNATION

A NUMBER of religions believe that when we die, we reincarnate and go back for another role of the dice at this human experience on Earth. Regardless of a person's religious influences, belief in reincarnation is a very personal matter.

For others, it's not just a belief, but also an experience. Jim Tucker, a professor of psychiatry and neurobehavioral sciences at the University of Virginia, has been actively studying and testing children who report having memories of having lived a different life before. Tucker is just one of a team who, over the past 60 years, have studied over 2500 cases. As part of the study, they interview the child and their family, then assimilate any reliable information they can piece together about the previously reported life. The team then essentially tests the child and cross-reference their research to analyse the accuracy of the child's apparent memories.

Tucker's experience is that children born with these past memories begin referring to them as soon as they are able to start communicating with their families. However as the children get older, these memories would fade. There are so many reports of children having memories of a previous life that it is hard to dismiss as coincidence. It is worth noting that Tucker and his team have been very aware of, and look out for, the possibility that children may have pieced together information from things they could have seen in the media or things they've overheard. The supporting evidence of these being real memories from someone else's life is really compelling. Tucker has published a number of these cases in his recent books.

Tucker and his team have also explored the science of how this concept of reincarnation could actually be a reality. Guess what it comes back to? Quantum physics and specifically consciousness. This is the foundational building block around which all else is built. Consciousness comes first, then physical matter like the human body and brain. When a person passes away, their physical body perishes, but consciousness remains. It transcends death. Now their research hasn't been able to define whether or not it's a specific consciousness that goes on to inhabit a new body or perhaps a memory within the universal consciousness that is being accessed. All the research I have conducted for this book and my own understanding, suggests the latter. There is no separation, we all have this universal connection, there are just some elements that are more prevalent in our awareness. The more we train the muscle of our intuition, the more we can access the greater collective consciousness.

DEJA VU

This brings me to deja vu, the phenomenon of feeling like you're reliving an experience, and something that has fascinated me since I was a young child and first experienced it myself. Emile Boirac first introduced the term deja vu, which translates from french to mean, already seen. If you haven't consciously experienced deja vu, the best way to describe it is the sense that you have been in this exact moment before. For example, imagine you are travelling in an unfamiliar city, walking in a strange area and you get the most overwhelming sense that you've walked this street before. This is deja vu. At times, it can be intensely strong and even unsettling, but at others, the sense is so fleeting that you may even doubt you had the flash of awareness. This phenomenon feels like a fragment of a memory, not quite a distinct memory, but definitely something familiar. It is proposed that up to 80% of people experience deja vu at some time or another.

There are reportedly about 40 different theories about the cause of deja vu, both scientific and paranormal, however enough science has been conducted to all but dismiss the possibility of it being a form of prophecy or precognition (but we'll get to that).

One of the most prevalent theories is the concept of dual processing. When a person is exposed to an environment, all the different stimuli coming into the brain are processed as a singular event. The principle of dual processing proposes that there may be a split second delay in processing one piece of information, so that once it is received, the brain adds it to the previous information and perceives the event as having already happened.

Another concept is the holographic theory. As you perceive information in your surroundings, it triggers a memory. This draws on the principle that memories are stored in the brain as holograms (i.e. you only need one fragment of the whole to see the entire image). It is suggested that when you see something that triggers a familiarity, the brain accesses the information and populates the memory without identifying its original origin. They call this familiarity without recollection.

A further popular hypothesis to explain deja vu is the theory of divided attention. Like in dual processing, we understand the brain processes multiple stimuli all at once. What differentiates divided attention is that it is thought that if we are distracted and only focusing on a singular object, then the rest of the information in our environment continues to be processed but instead occurs below the level of conscious awareness. So the information has entered our brain without us knowing how. In turn, we perceive the memory, again, without knowing its origin.

These are just some of the theories, however scientists have taken it one step further and have found a way to study the brain activity of people experiencing deja vu. They have discovered the brain activity reflects a process of the frontal part of the brain checking through the brain's library of memories.

Based upon this information, many now believe that deja vu is actually the checking process sending up an alert of a memory error. One particular study proposed that deja vu is more prevalent in young people and declines as they age, which would match age-related memory function decline.

Scientific research appears to have found a simple, logical explanation by recreating the sensation of deja vu in a laboratory and then analysing the resultant brain wave states.

So, what has the research found?

Do you want the good news or the bad news first? Deja vu has been found to be a very common phenomenon in people suffering temporal lobe epilepsy, as well as several psychiatric disorders. But… experiencing it is not a differential for diagnosis, just because a person experiences deja vu, doesn't mean they have a greater likelihood of temporal lobe epilepsy or a psychiatric disorder. However, if we flip the connection on its head, when researchers looked at people with one of these diagnoses, there was a significantly higher likelihood of experiencing this phenomenon. The good news is, deja vu is just as prevalent in totally healthy people without any diagnoses. To the point that it has been classified into two groups; pathological and non-pathological (e.g. in the presence of a diagnosis, and in the absence).

Researchers have also explored the possibility of a genetic predisposition based on neurochemical modelling, but as yet, nothing appears to be substantiated.

Can you believe that researchers have gone as far as to identify the exact part of the brain that will elicit an experience of deja vu if it's electrically or chemically stimulated? Let us just take a moment to process the significance of this, scientists have used electrical probes to stimulate regions of the brain in awake and willing subjects, in order to study deja vu.

Yet all this research and varying theories, still don't explain why some people experience it so profoundly, and others hardly at all. We have plenty of seemingly valid theories and a fabulous grasp on the science of what can go on in the brain, but it still doesn't connect all the dots.

I'd like to tell you the story of Michael and his experience, perhaps the most significant incident that I've ever come across. On this particular day, Michael, who was in his early-mid 20s at the time, had the most intense sense of foreboding that he'd ever experienced. What makes this story really intriguing, is that only a few moments later, Michael stepped out to cross the road and was hit by a passing truck. People nearby at the time, report that he was actually flung across the road from the impact. The accident was so severe that he was clinically dead. Fortunately a passerby was able to resuscitate him and keep him stable until first responders arrived. I have known Michael most of my life and his near death experience shaped a lot of my awareness from a young age.

Getting back to the subject, surely this profound moment wasn't just the brain cross-referencing a false memory, or a case of divided attention or dual processing. The fact that it preceded one of the most consequential moments of Michael's life is too much of a coincidence for me to merely dismiss it. I grew up knowing about this experience and personally always felt that deja vu was an intuitive hit. As a child I even believed that at night our dreams were rehearsals for the day ahead. I believed that in our dreams we would practice and rehearse everything we're about to experience. But unlike premonitions, we have no recollection of these events, unless it's something very significant in which case we'll have the sense of deja vu. Now as an adult, I don't believe this anymore, instead I've come to ponder if perhaps it is a form of survival intuition? A hit of awareness that we have no prior experience of how to process other than feeling like repetition of a lived experience.

I'm not ready to accept the scientist's theory as the complete answer, as I believe artificially creating the sensation of deja vu in a laboratory isn't the same as the random hits that spontaneously occur. What evidence there is appears to indicate that the sense of familiarity or a previous memory is as equally common pathologically as it is non pathologically. So for this topic, I can't give you a definitive answer, instead I leave you to ponder what deja vu means to you, and if perhaps it truly is part of our ability to tap into the greater Universal Consciousness.

6

THE INFINITE WISDOM OF THE BODY

"Everything you'll ever need to know is within you; the secrets of the universe are imprinted on the cells of your body."

The Way of The Peaceful Warrior - Dan Millman

The body has a wisdom that defies our rational brain.

Have you ever wondered how some people can seemingly punish their body and do everything "wrong" yet still appear to live long and healthy lives. Yet others who follow all the rules of health and appear to do everything "right" can still get sick and even suffer fatal illnesses

"Ding ding ", you hear the sound to indicate you have a new SMS on your phone. Is it a friend reaching out to say hi, or perhaps it is a family member reminding you to pick up groceries on the way home? Maybe it is a reminder that your electricity bill needs to be paid. Whatever the message, it will be important enough for you to at least look at the screen and register it.

Do you do the same when your body sends you a message? Do you even notice that your body has sent one? Now of course this miraculous but confusing paradigm of human existence would be a whole lot easier if we could receive clear signals from our body, telling us what it needs on a regular basis.

The thing is, it actually does, it's just that we're not fully attuned to its method of communication. Our body is always communicating with us. It is never not sending us messages to let us know what is going on and what it needs.

The body exists as layers. The innermost core is what we know, and regularly identify with as our physical human body. Just beyond this is our subtle body, an energetic space influencing and influenced by the physical body within and the universal energy without.

References to something matching the subtle body, can be traced as far back as 2nd century BC in recovered Taoist texts. Many other cultures have similarly referred to the energetic space just beyond the physical body. What western cultures now call the subtle body is an adaptation from a Sanskrit term and philosophy. It cannot be perceived by the bodily physical senses but our intuitive senses can. It is thought that the densest vibrations occur in our physical body. As we move outwards in our energetic field, and then into the Universal Consciousness, the vibrations continue to increase.

Everything is an interplay of energy exchange, and the human body is no different. It is always communicating with us. The dialogue starts as tiny whispers in the subtle body, fleeting little messages only perceivable to our intuitive senses. We can think of the energy of these messages as still vibrating relatively high so it is still able to reside in the energetic field. The more we ignore or reject these messages, like an irritable toddler being ignored, they get louder and more persistent. The messages become more apparent and the energy vibration lowers to a density that allows it to drop into the physical body. This is often when the majority of people first start to notice a little niggle or awareness that something isn't quite right.

Just because these subtle little messages start to become apparent here, doesn't mean we all necessarily stop to take notice, or perhaps even process its meaning. You see, these early little nudges are indicators that something in your current situation, isn't working for your higher good. At this point of the communication pathway, as human beings in general, we tend to either disregard the message or have a thought something akin to *"oh, that doesn't feel right, hopefully it will go away on its own"*. However, the intention of the message is for us to react a little more along the lines of "*that doesn't feel right. What is my body trying to tell me? What, if anything, do I need to change?*". You could also think of these signals as being like receiving a letter in the mail reminding you that your car registration is coming up for renewal. You know you'll get further reminders, but this is the first point of contact to hopefully propel you into action.

So what happens, if we're too busy being caught up in the treadmill of life, and totally disregard that first physical indicator? The message gets stronger, a little more insistent. The thought process that goes along with this level of message from the body may be a little more along the lines of *"That still hasn't gone away. Actually, come to think of it, it's gotten worse. Maybe I should try to make an appointment with the doctor."* Nudges from the body at this level are strong enough to trigger the thought of taking some form of action, however not enough for the thought to be held with conviction. If we return to the car insurance analogy, this is the point in which the insurance company sends you an SMS telling you that they've sent a letter, and your registration is now due.

Now of course, there's another level... We are all human beings after all, perfectly imperfect. Life is busy, we don't always have the time to stop what we're doing in that second, and take action from a thought that pops into our awareness. So, the body, in its infinite wisdom, keeps reminding us, getting more and more persistent. Until it reaches a point that the body sends a signal that can no longer be ignored, nor can it be forgotten in the busyness of day to day life. This time we stop and take notice *"Something's really wrong. I'm going to call the doctor right now"*. This is like the SMS reminding you that your car registration is due "today" and if you fail to register your car, you won't be covered if you chose to drive the car.

Depending on what type of person you are, it may or may not be surprising to know that many people still push on and avoid the message (some of this is deeply rooted in fear which I'll get to later). At this point in time, what does the body have to do to get you to take action? Usually it is some form of drastic health event. For example a stroke or a heart attack, something that is of consequence, like a diagnosis or a significant event takes place. It doesn't just need to be health related (although usually it is), that major catalyst may be something along the lines of losing a job, having a relationship breakdown or something happening in a career. Communication at this stage is a physical event of some sort.

I would love to say that if and when we take action at this point, we can always get back on the path of hopefully living our best life. It doesn't always happen that way. Our "best life" isn't always the one we think we're currently living.

Unfortunately, sometimes the body is meant to suffer illness. There are times that the messages need to anchor in the physical body as illness. Signals are not limited to being received by the individual, there are times that the messages are equally as important for the immediate family and even the greater collective. It is not for us to know the full meaning of each signal, but simply to appreciate and accept that every single message is somehow serving a higher purpose.

When we surrender to the love with which the messages have been sent, we have the greatest chance of receiving the intended outcome, whatever it is that serves our highest good. Please know that there is always a greater purpose.

The emotional pain and suffering we experience through illness are a result of pushing against the natural flow of events, the sense of helplessness that can occur, and the perceived injustice of it all. This is not so. You are not helpless, you are a divine part of this magnificent universe. The injustice you may feel is anchored in the 3D expectations of what your life should look like. Remember, we are not simply a three dimensional being, there is so much more to life (and beyond) than we can see with our limited senses.

So is there a way to avoid illness? Ultimately no, as it all occurs to serve us in some manner. But we can absolutely limit all unnecessary illness by attending to the needs of our energetic selves. By raising our vibration, we lift all stale dense, low vibrational energy, out of it's illness-causing state, and into a higher vibration where we have more freedom to process it.

Our vibration is the key to making the most of this human experience. We'll look at ways to tend to vibration and actively increase it in future chapters.

7

WHEN ILLNESS STRIKES

If we're all getting super enlightened, becoming in-tune with our intuition, evolving as a society, blah blah blah… then why are we still getting sick?

I don't know about you, but I've been watching "spiritual" people over the years and have taken a very keen interest in their progress. One of my biggest observations is that their physical bodies appear to get more fragile. Have you noticed this in the people around you, or perhaps even in yourself? I've certainly noticed it in myself, and as much as I've tried to rebel against this interesting trend, at this stage of my journey, I seem to be a helpless passenger along for the ride.

Firstly let me explain my observations, particularly from my own experience, and then I'll start unpacking what it's all about both from a practical physical perspective but also an energetic one.

Since being on my own path towards enlightenment (sorry, I just had to throw that cliche in there, in truth, I prefer to think of it as energetic discovery and freedom), I have noticed my own physical body has become much more sensitive. I once mindlessly ate anything, happily and easily functioned with very little sleep, and stress was an essential part of my day. Now, however, I no longer have that physical resilience. Perhaps this is simply a by-product of the number of birthdays I've celebrated, but I have seen this trend all too often amongst those who begin to honour and prioritise their energetic selves over their physical. The analogy that comes to mind is that we go from being cheap work-horse cars that go forever, to becoming high end luxury vehicles that require regular maintenance and only the best quality fuel.

Some may take the point of view that if our bodies are going to become more sensitive and require that much more love and care, why bother along the path of awakening? If we continue along with the luxury car analogy, why go through life, simply being, why not live it the very best way we know how? Keeping in mind that my "best" might be different to your "best", and even that may change from year to year, decade to decade, as we all continually learn and discover more and more.

One thing that is also at play in the noticeable increase in our physical sensitivity is that many spiritual seekers are constantly looking to improve their emotional and energetic state. Sometimes, they work so incredibly hard and overcome many demons along the way, that they are left with a sense of vulnerability in their human body. The only thing they can't yet conquer and control, is their physical form. Control is a very interesting thing as it goes against the very foundation of honouring ourselves as spiritual beings. In pursuit of control, we stop surrendering to our higher truth and anchor back into the muddiness of our perceived reality. Some have managed to gain so much control, mostly through hard work but also perhaps by not surrendering to the universe in the very fullest of senses. In the process of this diligence, the shadow of the control freak has the opportunity to creep in. The only thing left that can't be controlled is the physical body.

This becomes a layer of fear in the shadow self of a lot of spiritual seekers. They find that their physical body is their ultimate weakness. It is the one thing outside of their control, and it is the one thing that does not follow the rules of the path to enlightenment. This brings up an interesting point, in that we always have so much more to learn. If one sits in the energy of fear and tries to control the apparent vulnerability of their physical form, then perhaps they've by-passed one of the absolute fundamentals of the principle that we are energy first, physical second. If we truly master the energy, then we no longer need to try to exert control over the physical. This is the paradox of being a spiritual being having a human experience, we are still always learning. The day we don't need to learn any more is the day we graduate from Earth school. This is the day we no longer need to be in a human form and we move on. We cannot by-pass anything along the way.

So, if we can't control this physical form, then what's it all about? Our human body is always serving us. It is always serving our highest light. On our path to connection to the greater consciousness, our body is our best friend. It is life's greatest gift to live life in this human form.

The old cliche of everything happens for a reason is so incredibly true. This is also the case with illness. Some people like to call illness or any diminished health, to be an occasion of up-leveling. Whilst this can be true, I would argue that sometimes this is ultimately just our body serving our greater good. For example, it can be as simple as getting a cold and forcing us to stop. How often, no matter what we do, we still manage to find ourselves on the hamster wheel of life on this society's treadmill. When a simple virus like a cold comes along, it forces us to stop.

All too often, we don’t listen to, or see the signs and get caught up in the doing. When we are just doing, and not being, our body comes along and reminds us that we haven’t been listening. Perhaps we have missed all the subtle signs sitting in our consciousness, we have missed everything in the subtle body, and even the first signs in the physical body. So now, everything comes together to give us a huge big stop sign that can’t be ignored in the form of illness. So that little cold virus is actually a stop sign along the road of life.

This illness is a gift but, unfortunately in this society we don’t see things like illness or frailty as such. Our collective focus is on being superhuman and the apparent ability to be everything and then do everything for everyone around us. We are constantly pushing the boundaries of physical form, trying to go beyond our natural healthy limits. When we have to stop, or allow ourselves to show apparent weakness, then our physical bodies are seen as less than perfect. We become the creators of our own suffering. When we really struggle or become truly aware of the vulnerability of our human form, it can be a reminder that our bodies have a finite lifespan. So too can each passing birthday be a reminder of this. I think it's fair to say that most people struggle with the idea that our physical life span may perhaps be shorter than we’d like. Not only is this a confronting notion, but the uncertainty and fear of possible suffering compounds this as we age or face declining health.

I'm going to give you a scenario, admittedly about a fictitious person, but it matches scenarios I have seen played out countless times for real people who have either been very dear to me or people I have had the privilege of working with. In this instance, we'll call this person Susan. Susan is diagnosed with terminal cancer (yes I'm going straight for the big "one" because that's a big black fear cloud that is in the back of everyone's mind) and she and her family have been told that Susan has six months to live. What is the first thing Susan and her family do? They take on the approach and energy of fighting. It is a natural human reflex to want to fight against anything that negatively impacts our life. Susan goes into battle, waging war on the cancer. Ultimately this energy translates into fighting the body, and even going as far as cultivating hatred on the body. How can the physical body be strong enough to overcome the actual physical manifestation of something like cancer, as well as endure the intensity of current, modern treatments, in the energy of fighting, battling and hatred? Love and gratitude are the most powerful emotions we can cultivate. It is these emotions we need to cultivate in all aspects of life, even in the face of illness.

I don't say this flippantly as I've lost very, very dear loved ones to cancer and to see someone suffer is one of the most painful things. It is also horrendous to feel helpless, and it is something that no person should ever have to endure. I am deeply sorry if you have been in this situation, or will ever be in the future. It is my hope that you will remember this chapter with the intention that it is written. I hope you will have the opportunity to give love and compassion to your body or that of your loved one, and be able to tune in to ask what message the body is trying to send?

My wish is that you will be able to take this approach when any illness comes along? Is it here to tell you to stop continuing along your current path? Is it telling you to change your lifestyle? Is it telling you to spend more time with your loved ones? Is it here to remind you of the preciousness of life, of this human form? Sadly when we wage war and start to battle our body, we can switch into the negative and often become so consumed with what lies ahead, that we struggle to see the gift. Please, don't get me wrong, I totally get it that illness is an unwanted gift. But to continue with the gift metaphor, please don't try to send the gift back, throw it in the corner or try to set it on fire. Lovingly receive the gift with the intention with which it was sent, and don't forget to read the card and find out why you were sent the gift in the first place.

We are yet to unlock all the mysteries of the human form. It is something we will always be searching and seeking. As long as we are searching and seeking, we will still be constantly confronted with the mysteries of illness, so please rather than fighting and battling when illness lands on your doorstep, take the time to look at what is within. I understand that sometimes it is hard to go searching for answers, sometimes we'd rather turn our back on it, ignore the problem and pretend it's not there because it's just too scary to face. Instead thank your body, thank it for the gift and give it what it needs right now. In the case of the scariest thing of all, terminal illness, please remember we are all going to die one-day. It is the only certainty of life, that one day it will end. This reality is also one of the biggest things that causes so much grief. What would it take to begin to accept that the cycle of life must eventually come to a close, and and sadly sometimes illness is the reason. Death is an inevitability for us all, please don't fear it. We do more harm in fear than in the actual illness. As I come towards the conclusion of this chapter please know that I send so much love, strength and compassion, and everything possible on earth, to anyone going through this emotionally intense time. I know severe illness doesn't seem fair, and I don't write this chapter flippantly. My intention is to help ease the pain and to remind you of how much of a difference perspective can make. Your body loves you and is always working in your favour, for your highest good.

What do we fear? Is it ourselves or our loved ones getting sick? Or is it actually having no control over our body and its propensity to get sick? The fundamental problem is fear. If we find a way to conquer the fear, we can live in a great deal more peace and comfort and view illness through a different lens. What would it take for you to see illness through a lens of love, compassion and acceptance.

And who knows one day we might uncover all the mysteries of this human experience. Perhaps at some stage in our lifetime, we'll discover all its secrets and alleviate all physical suffering. I'd love to be that person to unlock it all. I'll admit, that despite all the work I've done, I still have a very human ego, but I also know that that's not the reason I was born into this lifetime. My purpose is to help people get the most out of their humanness, to find the gift in being human. I hope that by reading this book, you come away with a greater appreciation of who you are and how you fit into this wonderful existence. We are all spiritual beings having a human experience. The quality of that experience depends totally on our perspective. Do you live in fear, or in gratitude? Do you live your life in absolute, total congruence through every part of your life?

There was a popular acronym doing the rounds recently, YOLO - you only live once. This might seem like quite the oxymoron when we talk about being spiritual, energetic beings first and foremost, and physical beings second. But we do only live once in this perfect combination of this specific physical body, with these physical surroundings, and the wonderful mix of loved ones that surround us. We all have a reason that we're here, right now. Make the most of it as best you know how.

8

GRADUATION DAY

"Death is not extinguishing the light, it is putting out the lamp because the dawn has come."

Rabindranath Tagore

DEATH CAN BE one of the most confronting and painful experiences of our human existence. However, the ultimate, heart-wrenching pain is not suffered by the deceased, rather the loved ones left behind, struggling to continue without the physical presence of the person who has passed.

There are lots of cliches that get used when a loved one passes…
"They are finally free"
"They live on in your heart"
"Your loved one is always with you"

The pain we experience after losing a loved one may influence our own perception of death and in some cases even result in significant fear around dying. This can be magnified if we fear our loved ones having to one day experience the same pain we felt in grief.

Death is an inevitable part of being human. And it is important we understand what it means for us as spiritual beings.

What is death? It is the end of our physical form continuing to function as a living, breathing conglomeration of cells. Depending on our end of life choices, a physical body may still remain. This body is made up of lifeless cells, masses of carbon molecules ready to take on a new function, to return to Earth. This is all part of something called the carbon cycle.

Our energy however continues to exist, no longer tied to a physical form. Death is merely the end of the human being, not the spiritual. If you have ever experienced the pain (or the divine privilege depending on your point of view and connection to the deceased) of being in the presence of someone when they pass, they no longer "feel" like that person. There is a tangible essence of a human being that we don't notice until it is no longer present. That energetic spark that animates the mass of cells that is our human body, does not disappear. It still exists, but it is no longer a separate and tangible thing, it is omnipresent, everywhere and in everything.

Your loved ones never leave you. Once free from the perceived confines of their human body, they are free to exist in pure energetic magnificence. By all accounts, it sounds like bliss beyond anything we can fathom. I'm in no hurry to get there, this human experience has so much to offer, but I'm not afraid of death either. It is fair to say that

I even look forward to the day the time is right for me to discover for myself just how amazing it is.

Right now, we are in Earth School, doing the hard work, learning the lessons we need to learn, and equally teaching the lessons to those around us. Some days are really hard, and it's ok to find them challenging. We're here to be human, not perfect.. We are all already perfect in all our imperfections. Perfectly imperfect!

Your heart and your soul are the direct line to Universal Consciousness, and you will hear this as your intuition. It will never guide you wrong. Even if you do get something "wrong", who's to say it wasn't a necessary step to guide you to your next step along the journey of Earth School. Thereby making that "wrong" decision, absolutely perfectly right for the lessons you need to learn right now. There are no wrong decisions, just detours. One of the key tasks we are here to learn is to surrender to and trust in the messiness of living life as a human being.

The day we have learnt all the lessons we are here to experience, is the day it's time to graduate Earth School and complete our experience of being human.

9

IT'S A VIBE

NOT LONG AGO, my gorgeous teenage niece started saying it's "a vibe" in response to everything. We had a little chuckle at the time, of course not at her, but at society and how words and phrases become a dominant part of pop culture. I cringe to think back to what we always said when I was younger - Teenage Mutant Ninja Turtles was a huge hit so "cowabunga" made a few appearances. I definitely prefer "It's a vibe". It may not be in the context that my niece and her generation intended, but everything truly is a vibe. Everything is a wonderfully intricate, interconnected vibration of energy.

As we learnt in the infinite wisdom of the body, we as human beings exist in a number of layers of increasingly higher vibration as we move outwards from our physical form.

If you recall, in earlier chapters we discussed that raising our vibration is the key to finding balance between our energetic and physical body. I don't like to refer to these as though they are separate entities, rather layers of the same, so perhaps it's better to think of it in terms of increasing your vibration is the key to minimising the separation between your energetic self and physical body.

So how do we raise our vibration? This is a very individual thing and is ultimately anything that makes you feel absolutely amazing. For some it may be singing or dancing, spending time outdoors in nature, exercising, chanting, catching up with friends, being with loved ones, meditating, reading, writing, cooking, drawing... As you can see, the list is full of things that make your heart sing from its very core.

Other things that help raise your vibration are easier to think of in terms of being the opposite of things that lower it. Eating clean, healthy foods is one way, as making poor food choices lower your vibration. So too is getting enough sleep, as minimal sleep also causes a lowering..

The people you chose to be around are pivotal to your vibration, not just from the point of view of how they make you feel in your physical body, but also how their energy impacts yours. Energy exists in waves (much like waves in the ocean). When we talk about vibration, we are referring to both the distance between the peak of each wave as well as the height of the wave. The higher the vibration, the higher and more frequent the wave. Have you ever seen two different sets of waves come together, either from opposing sides of a headland or around a sand bank, if the waves come in matching or in unison, they amplify each other, thus raising their collective vibration. However if they merge out of sync or totally mismatched, they tend to cancel each other out. This also happens with people and is why it is important to be aware of the vibration of others around you. If this is the first time you have come across this concept, I understand it is a huge jump to be able to sense or get a measure of someone else's vibration. The best place to start is to take notice of how you feel around different people, both whilst you're with them and immediately after you leave their presence.

THE CHAKRAS

In addition to this overall concept of energy and vibration, our physical body is thought to be made up of a number of chakras which are vibrating energy centres. Exploring this world is like opening Pandora's Box, as there are so many chakras, on so many levels. It would easily fill a book in its own right. So for simplicity, we'll stick to what they are, and the main ones found in and around the body.

The word Chakra comes from the Sanskrit word for wheel and it is used to describe the spinning disks that are our energy centres. These energy centres correlate to major nerve centres and organ junctions along the autonomic nervous system.

The first recorded mention of chakras was in ancient Hindu religious texts dating back as far as 1500-1000 BC (just a point of interest here, these ancient texts are called the Vedas and are considered the oldest religious texts in history). There is also an overlap in the fundamental principles with that of meridians in Chinese medicine. However a great analogy of how they fit in the greater scheme of the body is to think of meridians as train lines throughout the body and the chakras are the stations.

The human body has seven major chakras starting at the base of the spine and then regularly spaced up and along the spinal column, with the highest being just on or above the crown of the head. The Base/Root chakra is the lowest vibration of all. As we move up the spinal column, they steadily increase in vibrational frequency, all the way to the Crown at the very top of the skull. The chakras are known by several names, primarily the Sanskrit name and a descriptive name that correlates to the physical location along the body.

Each is an energy centre that acts as a portal between the physical body, subtle body and Universal Consciousness. No matter which modality you follow, it is hard to deny the existence of these energy centres. In Chinese Taoist philosophy it is the Dantien, The Cabala refers to the Tree of Life and even Hawaiin Hua refer to Auw centres. Physiologically, the locations of the chakras correlate with significant organs and nerve plexuses.

The chakras each have a lesson to be overcome in order to live a balanced and open life. When energy is stuck in a particular location certain organs are thought to be affected as well as the emergence of possible mental and emotional issues. I am deeply grateful to the work of Carolyn Myss and the wisdom she shares in *Anatomy of The Spirit.* Myss has paved the way for us to view chakras through the lens of so much more than simply physical manifestations in the body.

BASE/ROOT - *MULADHARA*

Translation - "Root support"

Location - At the tailbone, base of the spine

Colour - Red

Lesson - "All is one"

Organs - physical body support, base of spine, legs & bones, feet, rectum and immune system.

Mental and emotional issues - physical family and group safety and security.

Physical dysfunctions - chronic lower back pain, sciatica, varicose veins, rectal tumours/cancers, depression, immune related disorders, panic attacks & anxiety.

Issues surrounding physical security and stability, family and community may arise here. Specifically the ability to stand up for self and to provide for life's necessities. Our ability to feel comfortable within our family and social settings lies in the base chakra. As such, it is the place where the fear of rejection stems from and where we hold feelings of humiliation.

SACRAL - *SVADHISTHANA*

Translation - "Her special abode"

Location - Along the spine, between the belly button and pubic bone

Colour - Orange

Lesson - "Honour one another"

Organs - sexual organs, large intestine, lower vertebrae, pelvis, appendix, bladder, hip area

Mental and emotional issues - blame and guilt, money and sex, power and control, ethics and honour in relationships

Physical dysfunctions - chronic lower back pain, sciatica, ob/gyn problems, pelvic/lower back pain, sexual potency, urinary problems, infertility

This is the space of all things related to sexuality, pleasure and creativity. Here we may see issues with relationships, and connection to others. The creative aspect of this chakra encompasses recreational as well as professional and has ties with our financial stability. This is also the space where we hold shame.

We need to live with honour and integrity in order to live in full health from this energy centre.

SOLAR PLEXUS - *MANIPURA*

Translation - "City of the shining jewel"

Location - Top of the abdomen, just below where the ribs meet

Colour - Yellow

Lesson - "Honour oneself"

Organs - abdomen, stomach, upper intestines, liver, gallbladder, kidney, pancreas, adrenal glands, spleen, middle spine

Mental and emotional issues - trust, fear and intimidation, self-esteem, self-confidence and self-respect, care of oneself and others, responsibility for making decisions, sensitivity to criticism, personal honour

Physical dysfunctions - arthritis, gastric and duodenal ulcers, colon/intestinal problems, pancreatitis and diabetes, both chronic and acute indigestion, eating disorders, liver dysfunction, hepatitis, adrenal dysfunction

The solar plexus is the home of our ego, personality and self esteem. It is also the foundation of our personal power and inner fire. This is where we find our survival intuition and the origin of the term "gut instinct". Problems arise here when we are unable to accept responsibility for the person we have become. Balance is found when we find appreciation and gratitude for the gift of our existence and surrender control to the Universal Consciousness.

HEART - *ANAHATA*

Translation - "Not stuck"

Location - Centre of the chest, in line with and just above the anatomical heart

Colour - Green

Lesson - "Love is the divine power"

Organs - heart and circulatory system, lungs, shoulders and arms, ribs and breasts, diaphragm, thymus gland

Mental and emotional issues - love and hatred, resentment and bitterness, grief and anger, self-centredness, loneliness and commitment, forgiveness and compassion, hope and trust.

Physical dysfunction - congestive heart failure, myocardial infarct, mechanical and functional problems in the heart, asthma and allergy, bronchial pneumonia, upper back, shoulder and breast cancer, thymus.

Anahata is the first of our internal power chakras. Symbolically it can also be thought of as the portal to our subconscious. This is the centre of our chakras and our connection to Universal Consciousness. Here love is the ultimate power and requires a balance between loving others as well as ourselves. Compassion and forgiveness are also key elements to harmony in this space. Any healing here requires.

THROAT - *VISHUDDHA*

Translation - "Purified"

Location - At the level of the throat

Colour - Blue

Lesson - "Thy will be done"

Organs - throat, thyroid, trachea, neck, upper vertebrae, mouth, teeth and gums, tongue, oesophagus, parathyroid and thyroid glands, hypothalamus

Mental and emotional issues - choice and strength of will, personal expression, following one's dreams and desires, using personal power to create, addiction, judgement and criticism, faith and knowledge and the capacity to make decisions

Physical dysfunction - raspy throat, chronic sore throat, throat cancer, mouth ulcers, gum and teeth problems, jaw problems, scoliosis, laryngitis, swollen glands, hypo- and hyperthyroidism

Similar to the solar plexus chakra, Vishudda requires us to surrender to Universal Consciousness and the divine will. Here we look at how we communicate and our ability to speak our truth. Listening is also a key component here. Just as we speak our own truth, so too must we listen and respect that of others. The religious act of confession is symbolic in this energy centre as it represents a clearing of anything that is not of our highest good.

THIRD EYE - *AJNA*

Translation - "Command" or "the qualified absolute"

Location -Above and between the eyes, in line with the spinal column

Colour - Indigo

Lesson - "Seek only the truth"

Organs -brain, nervous system, eyes and ears, nose, pineal gland, pituitary gland

Mental and emotional issues - truth, intellectual abilities, feelings of adequacy, openness to the ideas of others, ability to learn from experience, emotional intelligence

Physical dysfunction - brain tumour, brain haemorrhage and stroke, neurological disturbances blindness, deafness, full spinal difficulties, learning disabilities, seizures.

This centre represents the lessons related to mind, intuition, insight and wisdom. Here we find the centre of our perception and how we see the world. The overall energetic message here is of asking "why" and constantly looking within ourselves to seek and know more than we did yesterday. If you are reading this book, it is fair to say your *ajna* is very much guiding you right now and there is a great likelihood that you have already been doing quite a bit of work in this energy centre. At this point in time, I'd love for you to reflect on what you've already read about the five previous chakras and the crown that you are yet to read, and to remember the fundamental principle of this human experience - balance.

Many people like to use the word mystic - it is thought that the work of the mystic is to navigate and balance our two states of being - human and spiritual/energetic. This chakra requires us to find the balance in the truth and integrity of our beliefs. To let go of perceptions and beliefs that are not in alignment with our higher good.When we act against the energy of this centre, we close and block the energy that allows our deeper truth to arise.

CROWN - *SAHASRARA*

Translation - "The unqualified absolute"

Location - At the top of the skull, inline with the spinal column*Colour* - Violet or white

Lesson -"Live in the present moment"

Organs - muscular system, skeletal system, skin, pineal gland

Mental and emotional issues - ability to trust life, values, ethics and courage, humanitarianism, selflessness, ability to see the larger pattern, faith and inspiration, spirituality and devotion.

Physical dysfunction - energy disorders, depression, chronic exhaustion that is not linked to a physical cause/disorder, extreme sensitivities to light and sound, memory?

Our spiritual needs take precedence over our physical needs. Our physical existence is temporary yet our spiritual is eternal. In eternity, we understand the power of the present moment. Our obsession with time and its passing is all an illusion. Imbalance here occurs when we dwell too long in the past or worry too much about the future. Eckhart Tolle famously says, all we have is this ever present moment. WE create the imbalances that interfere with our ability to perceive and then receive information from the Universal Consciousness. The guidance will not make sense and may even be missed if we focus too long reliving the past or looking forward to the future.

This is the centre that returns us to oneness to be spiritual beings having a human experience.

REFLECTIONS

Although the location of the chakras are described by organs and landmarks of the body that are along the front of the body, the chakras specifically lie along the spine, this is because they follow the line of kundalini energy.

You may have noticed a theme as we ascend through each of the energy centres. Starting at the Base, it is very self-centric and related to those around us (much as a child is born into the world and depends on their parents for their existence). Continuing up, we mature into our sexuality, then our personal identity, we experience the true depths of love and truly find ourselves. In that we then find our voice and come to a place of finding our inner truth and following the voice of something deeper. The next stage of maturity takes us to our wisdom and how we see the world. We then come into "oneness" - we return to the Universal Consciousness. Some texts refer to the crown chakra as violet in colour, others as pure white, white being the representation of universal consciousness.

You may have also noticed that the colours mentioned, perfectly correlate to the colours of the rainbow. These are the colours of the visible light spectrum and what appear when white light is refracted. Let me pose this possibility for you. Consciousness is sometimes referred to as "the light" and many people who've had near death experiences speak of seeing a beautiful white light. We are a physical representation of this Our physical form refracts and separates the white light into the rainbow spectrum of colours. One of the key lessons in our human life is to overcome our belief in separation. Our physicality is a manifestation of our belief in separation and as such is a refraction of our higher truth.

The earth on which we live also has its own vibration, called Schumann Resonance. It is very interesting to layer the Schumann Resonances recorded with significant events throughout life's history. It is equally fascinating to look at the bigger picture of how this resonance has changed, especially through the lens of seeing the world awaken.

SCHUMANN RESONANCE

The Earth is a giant electrical circuit and the atmosphere, a weak conductor. The ionosphere is a section of the upper atmosphere that separates Earth and the lower atmosphere from the vacuum of space. This ionosphere traps electromagnetic radiation that results in an underlying energy frequency. This is the Schumann Resonance, named after Winfried Otto Schumann who discovered this phenomenon in the 1950s. It has been suggested that human beings are affected by Schumann Resonance, just as we in turn as a collective, equally influence it. The overarching theme of energy is that its waves are infinite, they are everywhere and in everything, in a constant state of flux, impacting each other. By this reckoning it stands to reason that our collective consciousness influences the Earth's energy which in turn also affects us. HeartMath Institute are currently mapping data relating to this and have noticed a correlation between frequency of Schumann Resonance and the number of significant major events that occur in society. They are also further exploring a number of different factors to look at the correlation between the Earth's frequency and that of human beings. I look forward to following their progress to see what they discover over time.

10

HO'OPONOPONO

THE HO'OPONOPONO IS a traditional Hawaiian healing prayer cultivating reconciliation and forgiveness.

"Ho'oponopono" is defined in the Hawaiian Dictionary as:

(a) "To put to rights; to put in order or shape, correct, revise, adjust, amend, regulate, arrange, rectify, tidy up make orderly or neat, administer, superintend, supervise, manage, edit, work carefully or neatly; to make ready, as canoemen preparing to catch a wave."

(b) "Mental cleansing: family conferences in which relationships were set right (ho'oponopono) through prayer, discussion, confession, repentance, and mutual restitution and forgiveness."

The foundations of the Ho'oponopono are centred around the polynesian belief that all illness stems from the energy created by wrongdoing, either by the individual or by a closely connected family member. It is believed that the practice of Ho'oponopono corrects the wrongdoing and seeks forgiveness and atonement.

Traditionally it was practiced by elders and traditional healers to rehabilitate someone after committing a sin/crime, or in the case of illness, to seek to correct the kharma of whatever sin is believed to have caused the illness.

It is also believed that the practice of Ho'oponopono is essential to maintain healthy relationships with both family, friends and with one's God or gods.

All modern reports have been passed down from Hawaiian elders. Morrnah Simeona, a kahuna lapa'au (healing priest) is credited with adapting the traditional family based practice of ho'oponopono to a broader approach that suits modern society that is currently used widely today. It has been adapted and embraced as an energetic tool for everyone seeking to restore balance and harmony.

Ho'oponopono is incredibly, incredibly powerful, and is a seemingly simple phrase of four mantras, repeated as often as needed, from the heart. As each mantra is recited, one cultivates and channels deep feelings of forgiveness, love and gratitude. These feelings and intentions can be directed towards another person or within. But traditionally this is a self practice to heal and release negative energy within. As we have covered in earlier chapters, by changing our own energy, we in turn, influence the energy of everything around us.

I am sorry
Please forgive me
Thank you
I love you

I've chosen to include the Ho'oponopono in this book, not only for its beautiful history but also because of my own personal experience with the practice.

In moments that I get too deeply anchored in this projection of reality, when I notice feelings of frustration, resentment, anger and lack of tolerance (yes, despite all the deep inner work, I've done, I still have these emotions, I am human, and my humanness is my greatest gift to this world), I come back to the Ho'oponopono. These emotions are very natural and it is commonplace to feel these in this balancing act of being human. Please don't ever feel that these emotions should be avoided or suppressed. However frustration, resentment, anger and intolerance etc are still unhealthy emotions and the negative energy that surrounds them is something we should aim to quickly acknowledge and work through. I'm far from perfect and there've been plenty of times I have permitted myself to wallow in these feelings. I still have moments of getting caught up without realising how low I've allowed my energy to slip and let these emotions get overwhelmingly strong.

Ho'oponopono is my favourite tool to get myself and the overall energy back on track. It is particularly useful when there has been a difficult conversation or a misunderstanding. I also find it incredibly powerful for self forgiveness.

I am sorry
Please forgive me
Thank you
I love you

In one particular instance, I was avoiding someone because I was dreading a direct confrontation over an ongoing dispute. This particular day, I knew I had no choice but to face them. In the lead up, I practiced the Ho'oponopono over and over again. At first, they were just words, I had trouble connecting to the emotion of them. Over and over and over. With each cycle, I felt my feelings and emotions begin to shift, slowly at first then stronger and stronger. I started out reciting the words with the other person in mind, apologising and asking their permission for the way I was feeling and then sending them love and gratitude. Slowly I started to feel forgiveness and understanding for their point of view, love and compassion for the person, and a deep gratitude for having them in my life. The more I repeated the Ho'oponopono, the stronger the feelings became and I found myself no longer seeing the other person as separate, they became an extension of me, and I them. With each repetition of the Ho'oponopono I was apologising to myself, to them, to the Universal Consciousness. Equally, I was expressing love and gratitude the same. By the time I had to face this person, I was looking forward to seeing them and the encounter was great. So great in fact that the dispute dissolved, we were in agreement about the matter, and they even offered to assist me with another matter. It was a very surreal experience. This instance highlighted the power of the practice and that it isn't necessary to have mastered the art of going within and trying to do the work on ourselves. When we truly surrender to the process, we connect to the universal one mind and the individual becomes irrelevant. We are one. What we do for one, we do for all.

There is a legend about a man known as Dr Ihaleakala Hew Len (a student of Morrnah Simeona). The legend states that he was able to "cure" every single patient in the criminally insane ward in the State Hospital of Hawaii. What makes this story even more remarkable is that he is reported to have done so without ever seeing any patient face to face. Instead he set up an office within the hospital where he could go over patient files. As he did so, he practiced the Ho'oponopono himself, cultivating the emotions within the words. Within four years, he was reported to have healed all the patients as well as significantly improved the health and wellbeing of all the staff.

TO WORK WITH HO'OPONOPONO YOURSELF

The Ho'oponopono is so powerful and something you can start working with right now. There is no special guide or teacher needed for you to get started. These are simply phrases, 11 deeply emotive words and a practice that can change lives.

To begin,

I am sorry
This statement is an offering of acceptance of responsibility for any wrong that may have occurred. There does not need to be an obvious event to be sorry for, it is simply acceptance that you are responsible for your own reality. There is also great humility in accepting that you have wronged another.

Please forgive me
Here the request is made to the universe irrespective of whether there has been a consciously made wrong. This request of forgiveness is believed to cleanse negativity.

Thank you
This expression of gratitude encompasses appreciation for forgiveness, for the opportunity for change and growth and for who you are and where you are right now.

I love you
The three most powerful words in the English language. This love is an absolute expression of purity from your heart, offered to one another, to self and to the universal consciousness. Sometimes this love may not come easily, but by reciting the Ho'oponopono over and over, the easier it is to connect with this emotion.

The more the Ho'oponopono is repeated, over and over again, the more the power of the words, amplifies. With each repetition the emotions and energy building and all other thoughts and distractions fall away. The power and magnitude of this practice has often brought me to tears of overwhelming love and gratitude for the preciousness of life.

If there is one thing you take away and practice from this book, I would highly recommend it is the Ho'oponopono. The act of surrendering yourself to the admission of fault, the vulnerability and accountability of asking for forgiveness underpin the ultimate acceptance of the truth that you create your own reality. Finding gratitude and love in this process is the most powerful thing to begin to start to change your reality.

Dr Hew Len refers to the Ho'oponopono as cleaning, we clean to get ourselves back to a state of zero in order to hear the divine/universal mind. In one particular interview, he is asked what it would be like if we were all "cleaning". He answers "The world would come to peace". Imagine if this simple practice were to be the answer to world peace?

11

MEDITATION

"Quiet the mind and the soul will speak"

Ma Jaya Sati Bhagavati

WHAT DO you think of when the word meditation is mentioned? It all depends on what prior exposure you've had to meditation. For my kids, meditation is an "annoying thing we have to do before art class". For my husband, it means sitting on his meditation cushion, cross legged, in the same place every day. For others, it may mean putting on headphones to listen to a guided meditation, or going to a weekly class with a meditation teacher. For me, it means a multitude of things, but mostly my sacred time to go within. Some days it's really uncomfortable, others it feels like nothing much, simply quiet rest time, and others it can be life changing when deep, deep awareness drops in.

Sometimes when I'm rushing my typing, I accidentally type mediation. Funnily enough, there's a good reason these words are so similar; meditation is like a mediation between your mind and your soul. Sometimes there's a bit of back and forth, a little anxiety and tension, and an overarching sense of dominance from your mind, but ultimately the soul also needs to be given its time to speak also, allowing space for balance to be found between the mind and the soul.

There are a multitude of types and ways to meditate. I've included a brief list of the types, what sets them apart from other forms of meditation and their specific benefits.

> *"Meditation is a process of lightening up, of trusting the basic goodness of what we have and who we are, and of realising that any wisdom that exists, exists in what we already have. We can lead our life so as to become more awake to who we are and what we're doing rather than trying to improve or change or get rid of who we are or what we're doing. The key is to wake up, to become more alert, more inquisitive and curious about ourselves."*
>
> Pema Chodron

MINDFULNESS MEDITATION

This is perhaps one of the most common and popular forms of meditation, based on principles of Buddhist teachings. The goal is to pay attention to your thoughts as they come to you, observe the thoughts and allow them to pass without attachment or judgement. This is a practice that doesn't rely on a guide or anyone or anything external and can be practiced anywhere, anytime.

I often struggle with this form of meditation, and don't easily let thoughts pass through my mind. I find myself wanting to reach for a pen and paper to make note of them before they pass through and be forgotten. Others I know find this amazing for quieting their frenetic minds.

SPIRITUAL MEDITATION

Interestingly, when we look at spiritual meditation, it's known more commonly as prayer. Yes, prayer is a form of mediation. This style has its foundations in Christianity, Hinduism and Daoism - although I'm sure there are a few religious leaders who would be horrified to have the sacredness of prayer, described as mediation. Here is a perfect example of how words are loaded and hold meaning for people based on their past experience. Prayer is meditation.

Some like to enhance their spiritual meditation experience with trinkets such as essential oils, incense or crystals. We'll talk more about crystals later, but ultimately the intention of any form of mediation is about the experience within and doesn't rely on any external stimulus. It is, however, very valid that people may use rituals and props associated with these to prepare themselves for meditation.

FOCUSED MEDITATION

As the name suggests, focussed meditation involves holding focus on a specific thing for a prolonged period. This can be done through any of the senses, such as focused breath, an image, a thought/feeling, mantra (a saying repeated over) or a sound.

Mala beads are often used during focused meditation to help anchor the thought and keep count of repetitions if a certain number of mantras are to be repeated. I speak more about malas in mantra meditation.

Rosary beads are quite different to mala beads, in both their origin and design but the premise of focus and intention combined with deep connection within, are identical. Each bead on the rosary has a purpose and is a prompt for a phrase and intention.

Interestingly, I was speaking to a mature Catholic lady who was telling me of how she felt a miraculous healing had occurred in her life. She believed the pivotal factor in this was that she made the promise to herself (and to God) to spend time every single day, before her family awoke, to sit outside and perform the rosary. The more I asked her, the more it occurred to me that her religious devotion was incredibly similar to my daily meditation practice. I had always thought of formal religious practices as quite restrictive and sometimes loaded with fear, but this conversation was pivotal for me to see the connection. How you perform and what you call your practice is irrelevant, the key factor is the intention you set.

MOVEMENT MEDITATION

Like the previous form, this style of meditation involves specific movement to harness awareness in a specific way to switch off the thinking, reasoning mind. Yoga and Qigong are forms of exercise specifically designed to enter this meditative state. We can also access this state by doing any repetitive exercise such as walking, cycling (best not road cycling in busy traffic) or swimming. I was a competitive swimmer during my school years and was in the pool most days. To be able to keep following that black line at the bottom of the pool for hours on end certainly switches your brain into a different brain wave state.

Many people report that exercise is fantastic for their mental health, and I can certainly attest to this too from my own personal experience. This can be attributed both to the chemical endorphins that are released during exercise as well as the benefits of entering sustained periods of meditative brain wave states.

MANTRA MEDITATION

A mantra is a repetitive sound or phrase used to clear the mind. This form of meditation is evident in a number of faiths, including Buddhist and Hindi teachings.

Mantra meditation can be performed audibly or silently, it is not about the volume, but the focused awareness that occurs in the brain through repetition. Although in saying that, there is something incredibly moving about being part of a group chanting a mantra over and over again. The resonance that can be felt in the harmonious sounds of so many voices chanting together, with the beautiful energy of togetherness, is breathtaking.

Focussed awareness on a word or phrase can be easier for many, than trying to focus on the breath.

Mala and prayer beads are a popular way to keep track of progress through the number of mantras (this is also sometimes known as japa meditation). Physically touching the beads through meditation can be grounding. Malas are also traditionally designed to have 108 beads plus a guru bead, although there are malas designed with less. 108 is a sacred number in both buddhism and hinduism. The guru bead is designed as both a point of reference (counting begins at the first bead after the guru bead, and with each repetition of the mantra, the person moves the beads through their fingers to progress to the next bead) and also holds spiritual significance. The guru bead (or sometimes known as the "Meru"/mountain bead) is the teacher and holder of a person's prayers and intentions.

TRANSCENDENTAL MEDITATION

As opposed to mantra meditation, transcendental meditation is practised with a more specific mantra, silently to invoke a restful state of mind that moves beyond thinking. It has been said that transcendental meditation is the "purest, simplest and most effective form of meditation".

The mantra is repeated silently for 20 minutes, twice a day. There is no expectation of perfection and thoughts are allowed to come and go during the practice, each time, gently returning to the mantra.

Transcendental is also reported to be the most widely studied form of meditation. There is one particular study that found it to be statistically, upto three times more significant in its benefit to the human body and brain, than any other form of meditation.

It is my goal in writing this book, to present as much available evidence to allow the reader to reach their own informed opinion on each topic. This however is not possible when it comes to transcendental meditation. There is simply far too much information. So I'm prepared to call this one, definitive.

When people refer to transcendental meditation, it is possible they are referring to a specific organisation by the same name (often abbreviated to TM), who have registered this phrase. The practice has been so widely embraced and adopted that it is now its own movement. TM was introduced to the western world in the 1970s, however the practice of transcendental meditation has a long history having evolved from ancient vedic teachings.

PROGRESSIVE MEDITATION

As a child, I had a lot of trouble sleeping. I was anxious and put a lot of pressure on myself to achieve. When I hit late primary school and early high school, it was particularly so. I wanted to do my very best at everything and had trouble switching off my thinking brain at night. This would regularly then progress to panic that I was awake too late and wouldn't be able to get up early enough for swimming training the next morning, and would also be tired throughout the day. The more I worried about this, the more I would then worry about how I wouldn't be able to achieve all I had hoped.

I have never thought of my stepfather as New Age, yet he has been one of the biggest influences in opening up my mind to possibilities. He has never used "new age" or "metaphysical" terms with me, but now that I look back, I can see the foundations of what he was teaching me and exposing me to. One of those is meditation. When I had trouble sleeping, he would sit with me and guide me through a form of progressive mediation. It was a form of body scan where I would start at my feet, notice them, then allow them to fall asleep. Then I would move up to my ankles and do the same. And so on, all the way to the top of my head. I don't actually recall ever making it that far as the practice relaxed me so much and allowed me to switch out my analytical, stressed brain. I still use this practice regularly.

LOVING-KINDNESS MEDITATION

The principle of the loving-kindness meditation is to cultivate feelings of love and kindness and then send these feelings on to another, to a group of people or even to all living beings.

This practice strengthens feelings of compassion, kindness, and acceptance toward oneself and others. As well as sending out feelings and thoughts of love and kindness, one also opens themselves up to receiving in return.

Loving-kindness is particularly beneficial for those who have a tendency to struggle with feelings of anger and/or resentment.

GUIDED MEDITATION

All of the meditations listed can be self guided, i.e. you can sit (or walk in the case of movement meditation) in silence and solitude or they can be guided by someone else. Going to a regular class is a great way to create a regular ritual as well as be guided by someone experienced in meditation. Alternatively, guided meditations can easily be accessed from the comfort of home, or your preferred place to meditate (I personally love sitting on the headland, overlooking the beach) via meditation apps, recorded tracks, websites and even youtube. The key is to find what works for you. And to allow yourself freedom to play around with different options dependent on your needs at the time.

VISUALISATION MEDITATION

Finally, let's look at visualisation meditation. I may be a little biased, but I believe I've saved the best for last.

If you can visualise something, you can meditate on it.

To hold a visualisation, involving all the senses (or as many as you can) to bring the scene to life. For example, if you're visualising yourself on holiday by the ocean, you will imagine hearing the waves and the seagulls flying by, you may bring to mind the salty smell of the sea or of sunscreen. You might imagine the feel of the sun warming your skin or perhaps the breeze that often blows by the shore. You might also try to recall the taste of something you associate with holidaying by the ocean, like ice cream or fish and chips. Whatever vivid association you have with the scene, will further enhance the visualisation. This is of course, in addition to the imagery you bring to mind.

This practice is especially good for cultivating emotions such as peace, relaxation and calmness that may go hand in hand with the scene you visualise. It can also be a fantastic practice to aid goal setting. Visualisation meditation helps increase focus and motivation as well as being a form of mental rehearsal.

> *"Feel the feeling but don't become the emotion. Witness it. Allow it. Release it."*
>
> Crystal Andrus

WHAT HAPPENS TO THE BRAIN WHEN WE MEDITATE?

The human brain has the ability to change and adapt through our lives. Grey matter has been observed to thicken or shrink and synapses (the connections within the brain) can be created or can break down and disappear. This remarkable trait is called neuroplasticity. To simplify, imagine the brain has a plasticity/plasticine like element that we can mould and shape by the actions we take in day to day life.

There have been a number of studies exploring the physical changes that occur to the brain with meditating. There's one particular neuroscientist and her findings are well worth mentioning as she has thoroughly investigated both the effects of long term meditation as well as the changes that occur when first starting a new practice.

Sara Lazar (of the "Lazar Lab", a department within Harvard University) has proposed that regular mediation may delay or even prevent thinning of the frontal cortex (the part of the brain associated with memories) that naturally occurs as we age.

One of her studies of a group of regular meditators in their 40-50s, found that they had the same amount of grey matter in their frontal cortex as 20-30 year olds. This is a very interesting finding given that grey matter is associated with memory and is known to decline with age.

Another of Lazar's key studies was conducted with a group of non-meditators. In the study, she had the participants conduct an eight week course in which they performed mindfulness meditation activities (such as the ones described above in the types of meditations) for 30-40 minutes every single day.

At the end of the eight weeks, she used MRI to test the brains of the participants. Lazar found that brain volume increased in both the hippocampus (the area associated with learning, memory, emotional regulation, spatial awareness), as well as the temporoparietal junction (involved in empathy and compassion).

She also found that the amygdala (the area responsible for fight, flight, freeze response, and perception of threat) decreased in volume. Interestingly, the decrease in size of the amygdala directly correlated with a decrease in stress reported by the participants, even though their external environment did not change throughout the experiment (i.e. the only thing that changed was their brain).

These studies, in a nutshell, highlight that meditation not only improves the physical health and functionality of the brain, it also helps us feel noticeably happier.

If we simplify things and compare the brain to a muscle such a bicep, the more you exercise the bicep, the bigger it will grow and the heavier the weight it can lift. So too with the brain, the more we practice sustained exercises such as meditation, the more we teach our brain to be calm, relaxed, happy, compassionate, grateful, peaceful and emotionally regulated.

> *"Well-being has been found to be elevated when individuals are better able to sustain positive emotion; recover more quickly from negative experiences; engage in empathic and altruistic acts; and express high levels of mindfulness."*
>
> R. Davidson PhD & B. Schuyler PhD

WHAT'S WITH THE HAND POSES?

Have you ever seen people sit in the classic meditation pose with legs crossed and hands resting on knees with thumb and index finger touching? This hand pose is a mudra and is one of many.

A mudra is the term given for a symbolic or ritualistic gesture invoked during some meditation practices. Most are performed simply with the hands, but there are others that involve the entire body. Each finger in a hand mudra has a specific meaning, as do the body mudras.

The word mudra comes from the Sanskrit word for seal. They are thought to "seal" the energy and intention of the meditation.

There are far too many to delve into here, but ultimately, my point of view is that if a mudra feels comfortable for you or if you are an experienced meditator and find benefit in them, I have absolutely no right to say, don't do them. However, I raise the topic in this chapter because people are inquisitive and there is a lot of iconography of people performing mudras during meditation. But, you do not need to do any specific ritual or practice to gain benefit in your practice. If you have no energetic belief in mudras, it is my personal opinion that you will gain no further benefit by doing them. The most important thing is to do what is comfortable and what works for you.

FINAL THOUGHTS

> *"The goal of meditation isn't to control your thoughts, it's to stop letting them control you"*

The Age of Enlightenment

Ultimately, people have been meditating since before anyone can actually fathom. although the word meditation hasn't always been widely used. This was especially apparent in my own family. Whilst writing this book, the awareness dropped in that I was first introduced to meditation when my stepfather taught me progressive meditation to get to sleep at night.

When I had this realisation, I shared it with my mum and mentioned to her how amazed I was to realise that I'd been meditating since I was in primary school. She told me that she was taught the exact same process in prenatal classes whilst she was pregnant with me. However it was called a relaxation process at this time. So whatever you chose to call it, however you chose to practice, there is significant evidence to support the fact that you're going to be healthier and happier by introducing some form of meditation into your daily life.

12

LEARNING TO GO WITHIN

Knowing and embracing your true nature as a spiritual being having a human experience is not a passive process. It is not something you can sit back and just hope will click into place. That's not to say that you don't already have everything you need to know yourself as a spiritual being. Everything you need is already within you. You are all you need. You have everything you need. But how do you find it? How do you access it? This chapter is a summary and somewhat of a go to guide to show you how to access that deep divine truth within you, how to unlock your unlimited potential. And how to find the I in I Am.

Firstly, let's look at how you start your day. Do you have some form of devotional practice that anchors you back into your oneness? Or do you wake up each morning and step straight into the treadmill of human doing?

helps you tap into the 5D, into consciousness. Creating a morning devotional practice is incredibly important because it connects you from the start of your day to consciousness, into the 5D.

It is important to set yourself up first thing in the morning, before you get a chance to get plugged into the 3D projection of reality. Before you get distracted and consumed in the 3D hologram of doing rather than being. I am the first to admit this isn't always easy, some days it's incredibly hard, it takes discipline.

Spiritual devotion is not a chore. It is however a disciplined practice that increases your higher good, raises your vibration and strengthens your connection to the universal consciousness. Some people instead prefer to call their spiritual devotion practice, blissipline. I love this term as it describes spiritual devotion so well. It is a blissful discipline. Blissipline.

Now, I can imagine you are thinking that I'm going to suggest you roll out your yoga mat or your meditation cushion and sit there for 30 minutes in absolute silence, perfection. I don't expect that, I rarely even manage that myself. Your blissipline practice is what works for you. It may even look different from day-to-day, depending on your needs, time and surroundings. All it needs to be something to take you out of your busy analytical mind and connect you back into your heart.

For some, exercise is a form of blissipline. However not all forms of exercise are suitable. Exercise needs to be done in a deep, heartfelt manner, connecting you within, rather than forms that punish the body, or are done purely for gratification of external physical results. Others may prefer reading. Reading books such as this one, words that act as a reminder that you are more than what you perceive in this reality. A blissipline can be meditation or prayer, it can be deep, thoughtful contemplation within. It can be walking. It can be listening to music, chanting, journaling, sketching. Your practice can be a multitude of things, the only key criteria for whatever you do to set up your practice is that it takes you somewhere deeper. It is yours to remind you of your true nature as a spiritual being having a human experience.

In addition to your morning practice, it is important to look at how best to maintain this deep, connected energy throughout the day. The first step is to look at what triggers or aspects of life lower your vibration and drop you out of the awareness that you are a spiritual being first and foremost? What provokes you or causes significant stress? What gets you caught up so much that you fall back into being dictated by your analytical brain?

These are key points to look at (and please remember it is totally and completely normal to have moments of falling back into your analytical brain throughout the day) to identify where you can add short little vibration top-ups to return you to your blissful state.

I've mentioned heart coherence in more depth elsewhere in this book but it is such a quick and simple, yet powerful practice that it is worth mentioning again. Whenever you need a quick little tool to instantly reconnect to your spiritual connection, you can't go wrong with heart coherence. As corny as it may sound, it can even be done when you are taking time out of your busy day to go to the bathroom. Take an extra 60 seconds in the bathroom to focus on your breath, draw your awareness to your heart space and foster feelings of joy, love and gratitude here. You can also easily do this when getting up from your work to get a glass of water. This practice is especially helpful to me during my days when I'm starting to feel stressed or overwhelmed. Whenever I take a minute or two for Heart

Coherence, I noticeably feel myself coming back to centre. I remember who and what I am and all the stresses and irritations of the human world no longer seem so intense.

Breathwork is another powerful tool that you can work with anywhere and anytime throughout the day. There are so many different forms of working with your breath, so I suggest that if this is something that feels right for you, try out a few different practices to see what is the best fit. One of the simplest and most common ways to work with your breath is calmly and deeply inhale, pause, then a slow, full exhale, again pause, then repeat a slow, calm, deep inhalation, pause, slow, full, exhale etc. continuing for as long as you wish. The key is to not go beyond the limit of your comfort but try to exhale slightly longer than you inhale. For example, a comfortable rhythm may be something like inhale for the count of six, pause, exhale for the count of eight, pause etc.

Music has been a constant companion for me during the writing of this book. Whenever I have sat down to write, I've had music in the background. Songs and music for me are incredibly powerful tools to connect to my emotions. You don't need to wait for your vibration to dip to "top it up" throughout the day. I find that playing songs I love, that make me feel amazing, keeps me joyful, light and happy. If you haven't already done so, I strongly recommend creating yourself a playlist of music that lights you up. If you need a little head start or some inspiration, I've created a public playlist on Spotify called Spiritual Being Human. Here you'll find my soul songs. These are the songs that deeply touch me. When I hear them I get goosebumps, with some I'm transported to somewhere totally different, so happy and joyful, and others bring tears to my eyes no matter how often I hear them.

Another great tool to surround yourself with are inspirational messages, cards or images. I'm not necessarily suggesting you plaster your wall with quotes and pictures, but a little card on your desk where you'll see it throughout the day, or create an uplifting image as the screensaver on your phone works wonders. For years I had the following quote as my screensaver. My friend had sent it to me and it resonated so deeply that no matter how many times I looked at my phone, it brought little bubbles of joy into my heart.

> *"You've always had the power my dear, you just had to learn it for yourself"*
>
> Glenda the Good Witch

There are so many wonderful little hacks and tools to draw on throughout the day. You are a spiritual being having a human experience, do whatever you need to do to remind yourself of that at every opportunity.

When I was in high school, the favourite saying of one of our principals was "*You don't plan to fail, you fail to plan*". I always thought this related to the administrative side of life, but now that I'm older, I can see that it is a valuable lesson for every part of your life, even topping up your vibration and reminding yourself that you are a spiritual being having a human experience. Take a little time to set yourself up to succeed, have everything organised and easy to access so that if you start to feel your vibration dip or you have something challenging to overcome, your tools will be easily accessible, ready to go so there's no excuse. Seriously, all it takes is a minute or two and you can lift your vibration and turn your energy around. Energy is everything so take special care of it.

13

RELIGION

I ALWAYS FELT this chapter would be an essential part of this book, from the very get go, I knew it was essential to include a chapter about religion. Faith and the understanding and acceptance that there is a force greater than we can perceive, is an integral part of spirituality.

Surprisingly however, this is the hardest chapter to write, because it is such a personal topic for everyone. Each of our own relationships with religion is unique and so deeply loaded with so many layers. So I ask as you read this chapter to come in with an open mind and heart. Please be prepared to view the world through other eyes. Understand that our societal conditioning influences the way we view religion and our past experiences impact how willing we are to embrace religion. Preconceived judgments may also arise here. I have tried to present the information more from a theological study point of view, interspersed with a few of my own personal observations. You have full permission to totally disagree with what I say. Some of it is my opinion, and that is the beauty of the era in which we live. As an Australian woman, I don't take it lightly that I have the freedom of speech to give my opinion on this topic. I deeply recognise that not everyone in this world has (or had) the freedom of speech, nor the freedom of choice regarding religion.

There are countless religions (estimates put the number somewhere in the vicinity of 10 00) in the world but as this is not a theological textbook, I won't be going into them all. If I have overlooked mentioning your religion or one that holds significance to you, I mean no offence. In this chapter, I have attempted to cover the most popular religions within the world as well as those that represent significant developments in the views of society.

I will also share my own personal belief, which may or may not resonate with you (therein lies the beauty of what this book represents, it is full of facts and research, but ultimately what you do with the information will depend solely on your own interpretation). I grew up in a Christian family (one side Anglican and the other Catholic), Whilst we did not attend church, respect for the christian God as well an overall respect for all religions in general, was an important part of our daily life. I was permitted the freedom to learn and question religions other than our own and came to realise that whilst there were some glaring differences, nearly all had a similar underlying theme, there is one God, and there have been many messengers or prophets sharing the wisdom of God. Again, I mean absolutely no disrespect if I'm oversimplifying things, but please keep an open mind as I further explain my own ponderings.

It has been my observation that in each era of civilisation, there has been a messenger or messengers around which a religion has been based. I believe that these were real physical human beings in every sense of the word but interpreted and then shared the wisdom of God in a manner appropriate to the era. No one was wrong, they're all right, and they were absolutely perfect for the era in which they were founded. Despite this, religion has been the great equaliser and divider of humanity. Wars have so often been started based on differing religious beliefs and the ability to accept or even consider a different perspective on life.

Over the years, some religions have changed and adapted slightly with society, whilst others have changed very little. Again, I want to reiterate that it is not my intention to cast judgement or skew the reader in any way. Religion is a deeply personal thing and can trigger extremely strong emotions depending on past experiences (either good or bad).

I'd like to propose the possibility that as society is awakening, we now have the freedom and inclination to explore our own understanding of and relationship with God. In history, religion was deeply sacred and only the very chosen few were deemed "special" enough to have a direct connection to God. Religions acted as a form of gatekeeper to any spiritual experience. For the most part, we are now however living in a much more open and accepting era. The vast majority of us are free to explore and nurture our own individual connection and are safe to question some of the doctrines in which we've been raised. We are ALL so very special and have every right to be prophets and talk directly to our own God.

So what is God? I recently read Mary Magdalene Revealed in which Meggan Watterson shares that Mary Magdalene referred to God as Good. This resonates deeply, as God is not a man as we've been led to believe. God is not a human being, God is so much more, an energetic intelligence that is the entire universe, a state of being within us all, a connection that defies explanation. It is the tiny little strings that instruct life, it is the energetic bonds between atoms, it is the feeling in our hearts, and the invisible ways they interact with others. It is everything, yet it is nothing tangibly physical. God simply, is.

> *"The 'second coming' of CHRIST is a transformation of human consciousness, a shift from Time to Presence, from Thinking to Pure consciousness, not the arrival of some man or woman."*
>
> Eckhart Tolle

I have since learnt that the foundations of the Bahá'í faith are similar to my own ponderings and observations and have even been echoed by spiritual thought leaders such as Eckhart Tolle.

So what actually constitutes a religion? A religion is a social-cultural system of beliefs and practices that connect society to a divine higher order. By definitions, a religion is a cultural system of beliefs, practices and ethics, and at times, was a societal structure in which to inspire a deeper meaning to life whilst also maintaining a level of structure and order to the population.

It is estimated that there are up to about 10 000 distinct religions practiced around the world. I find that number absolutely staggering. Especially considering about 84% of the world follow the four most common religions, Christianity, Islam, Hinduism and Buddhism.

Most religions give reason for the events we observe in our day to day life based on sacred narratives to result in an ultimate meaning to our existence.

Many have a single deity responsible for the very existence of our life and an ultimacy that comes at the end of our life. Within each religion is a set of beliefs or “rules” by which to live in accordance with the values set out by each religion's deity or deities. These are designed to guide its followers towards what each religion ascribes to be the best way of life.

All religions are based around a system of faith that requires followers to accept and believe in a higher order that, during their lifetime, they will not be able to prove or disprove. It is perhaps this element of faith that results in so much emotion around religion.

I am not a theologist and I have not formally studied religion. Here is a brief overview of each of the most common religions followed around the world as I understand their teachings. As I’m sure you can understand, to truly explore each religion could fill a book so I have briefly given an overview and any key points I felt are relevant to this book. I have also included two further religions, Bahá'í and Gnosticism. These two are fascinating to look at. One is a relatively young religion and the other one of the most ancient, yet they both serve to provide a deeper understanding from a more open and holistic perspective and help embrace broader concepts of spirituality and mysticism.

CHRISTIANITY

Christianity is currently considered the world's largest religion. It is proposed that there are about 2.4 billion Christians in the world. Christians believe that Jesus was the son of God. The Old Testament prophesied his birth and much of his life. The vast majority of Christianity is based on the teaching shared by Jesus as well as a number of gospels, Matthew, Mark, Luke and John. One of the key teachings is that Jesus lived and died in order to save humanity. After suffering during his crucification, he rose from the dead as a symbol of the eternal nature of spirit and forgiveness for all sins.

It is believed Christianity first emerged in the first century as part of Judaism. Judaism is another religion (practiced by people of Jewish faith) that has many similar foundations and principles as Christianity, although it is perhaps more appropriate to say Christianity has similarities to Judaism as Christianity emerged from Judaism after the Fall of Jerusalem in the year 70 AD.

Within Christianity there are several different branches, such as Catholicism and Protestantism as well as a number of different Orthodox churches specific to their region of origin.

Many aspects of life in the wider western world remain based around Christianity such as Christmas (the birth of Jesus), Easter (the death of Jesus) and even how we number the years on the calendar with the year 0 having been the year Jesus was born.

ISLAM

Islam is considered to be the world's second largest religion, with approximately one quarter of the world's population among its followers, muslims. The first historical records of Islam have been traced back to 600-700 AD.

The teachings of Islam are based around the prophet Muhammad, a messenger of God to whom the Quran was revealed during his lifetime. The Quran is considered to be the ultimate word of God and in it are teachings about all aspects of life.

HINDUISM

Hinduism is reported to be the world's third largest religion with over 1.25 billion followers (that equates to approximately 15% of the global population). Emerging somewhere around 500-200 BC.

Hinduism is a mixture of beliefs and practices which instruct how best to live life (dharma). Some of the core beliefs of Hinduism involve what are known as eternal duties; honesty, never hurting a living being, patience, self-restraint, virtue and compassion to name just a few. The overall aim of Hinduism is to live by these eternal duties in order to attain liberation from the cycle of death and rebirth.

Hinduism has multiple deities. It is recognised that there is one supreme creator of the universe, however each of the deities represent an important attribute of the Supreme Absolute. It is my understanding that acknowledging and paying reverence to multiple deities allows the opportunity to pay respect to the multitude of elements that contribute to making this wonderful world.

Yoga and meditation are core practices in Hinduism however they are now widely practiced around the world, sometimes with minimal connection to the foundational beliefs of the religion. One of the things that makes these practices so appealing to the wider population is the overall sense of peace that comes as a result of yoga and meditation. Perhaps this is due to the inner connection to our own soul that occurs, then leads to a deeper awareness to what Hindus call the supreme soul of which we are all part.

BUDDHISM

Buddhism originated in India sometime between the 6th and 4th century BC and then spread across much of Asia. Its original teachings are credited to Gautama Buddha although there are some that attribute Buddhism to having evolved from Hinduism. Currently it is reported that there are over 520 million followers of Buddhism (Buddhists) , this correlates to approximately 7% of the world's population.

There are several branches of Buddhism but at its core, it aims to overcome the suffering humanity experiences as a result of being blind to the truth of our existence. The beliefs and practices of buddhism are designed to guide one towards transcending their current perception of reality, and the finite cycle of life and death, towards ultimate Nirvana.

One of the main differences between Buddhism and Hinduism (as well as all other religions) is that Buddhists do not attribute reverence to any deity, rather they follow the teachings of leaders who have gone before them and attained the ultimate goal of enlightenment.

BAHÁ'Í

The Bahá'í Faith was established in the 19th century. It is believed that there is one single God, responsible for all that is. Throughout history, God has sent a number of prophets, who correlate to many of the teachers/founders of the world's religions (eg Buddha, Jesus and Muhammad).

Since then, the Baha'i Faith recognises that there have been two more recent prophets, Báb and Bahá'u'lláh who have shared this broader explanation of God and all other religions. The primary goal of Bahá'í is unification of world order and love and acceptance to all.

GNOSTICISM

In researching this chapter, I came across the religion of the Gnostics. Now this isn't to be confused with agnostics which is an uncertainty to the point of indifference about the existence of a God. Gnostic however is a religious movement that evolved around the first century. It has its foundations in Judaism and Christianity and believes in a god-like higher order to life. The focus of Gnosticism aimed to answer the fundamental questions of human existence; "Where did we come from?" "Where are we going?" and "Who are we here to be in this life?". Whilst these questions are much discussed philosophically, the Gnostic movement aimed to explain these from a religious and spiritual perspective.

Through the expansion of travel, information and cultural exchange all around the world, it has become customary for many people (those who are free to do so) to cherry pick from multiple religions to come to a place that aligns with their own morals, background and philosophical understanding of life. Ultimately, I propose that there is no right or wrong. Hopefully in this chapter you've been able to see that each religion, at its core, aims to instil the same values and morals in its followers and are all underpinned by the same foundational belief that there is a higher power and order to our existence as human beings.

We, as human beings, are not here by accident. Our existence is part of an overall hierarchy or order and meaning. Religion aims to remind us of this and provide us comfort and support in times of hardship.

14

THE BALANCING ACT

Living life as a spiritual being having a human experience is a balancing act. Our goal is to balance knowing that we are spiritual beings first and foremost with showing up and getting the most out of being human.

We live in a time and era where being the perfect "spiritual person" can be challenging and for many, very daunting. Comparisonitis (the ever present habit of comparing oneself with others, especially on social media) has led to so many people feeling that they're not enough. Not doing enough, not being enough. And just because a person is on a path of awakening spirituality does not mean that they are exempt from this. We see others out there meditating, eating perfectly, staying blissfully calm all the time and generally nailing the human experience.

It is not just to our peers that we have a tendency to look at and then compare ourselves. I'm sure anyone who has been exploring, seeking and learning along the spiritual/new age path can think of a handful of teachers and mentors to whom they aspire to emulate. These people we have inadvertently placed on a pedestal, seem to have it all, they have all the answers, eat the perfect foods, sustain healthy relationships, have all the right spiritual practices to keep them zen and even look the part. I'm sure you know the "look" I mean… you know the floaty fabrics (or yoga gear), big dangly earrings, long flowing hair, beads or big chunky crystals around their neck… My husband likes to call this look the "spiritual uniform". Unfortunately, when I try to dress like this, I end up looking like I'm playing dress-up in loose flowing pyjamas and unbrushed hair.

Just as a lot of leaders teach that reality is merely an illusion, so too is the perfect image these very same leaders project to the world. Perfection is an illusion. No one is perfect. No one has it all. We are all right here, right now, having this human experience for a reason, learning the lessons our humanity has to teach.

Living as a spiritual being having a human experience is about balance. It is finding harmony between the visceral realities of life and the deep wisdom of our spiritual truth. Life as a human being means experiencing the amazing spectrum of emotions available to us, it is the delicate intricacies of families and of interpersonal relationships, it is the pressures of society, the search for identity and security and the fragility of the human body.

Being human is not easy. Do not let anyone allow you to feel any less because you haven't mastered the art of being human. There is nothing to master. You are exactly where you are meant to be, exactly as you are, right now. Please allow this to be permission to forgive yourself for being anything less than your own perception of perfection. Chances are that there is someone out there, looking at you, wishing they were like you, doing such a fantastic job of being human.

I believe there is only one thing you truly need to be consciously making every effort to do, and that is to be a GOOD human being. By that I mean, doing right by others, and your best to do no wrong. Every person walking around on this Earth is here to learn their own lessons and is trying their best. Our energy influences that of everyone around us. For a moment, imagine an existence where we're all cruising around with the primary intention to be a good human being and to do no harm to anyone else. Tuning into this possibility for me, feels like Nirvana. Having the entire world existing from the heart centred emotion of love, focused on doing good by others, would truly change the world.

So, short of brainwashing the world's population via putting something in the water, by 5G or by immunising the world with a new genetically encrypted vaccine (see Conspiracy Theories), how do we ever have a chance of creating this reality? In all honesty, this may never come to fruition, but that shouldn't stop us trying to be a "good" human being.

By focusing on that one primary intention, we inherently raise our own vibration and tend to the needs of our own spiritual health. By doing this, we then positively influence everyone around us, little by little, our high vibes and good intentions may actually help change the world.

So how do we balance our spirituality with our human selves? We allow ourselves to be totally present in the experience of being human. We keep ourselves open to the lessons we are here to learn whilst remembering that we are spiritual beings first and foremost. Yes it is that simple.

However, implementing this philosophy is just another of the profound lessons we are here to learn.

Don Miguel Ruiz perhaps summed it up the best in his book The Four Agreements. Based on ancient Toltec wisdom, the book presents four simple principles by which to live;

Be Impeccable With Your Word
Say only what you mean and only use the power of your words with good intentions.

Don't Take Anything Personally
Nothing anyone ever does is because of you. It is because of them. Their choice, their history, their interpretation, and ultimately their lessons.

Don't Make Assumptions
Never assume what others think nor assume that they know what we are thinking. Always be clear and open with all communication.

Always Do Your Best
Your best is all you ever need to do, and if you happen to grow and learn more, then do better.

Living by these principles allows us the freedom to enjoy this human experience to its fullest, to freely experience the most powerful human emotions of joy, love and forgiveness, to not be encumbered by expectations caused by making assumptions and taking things personally, and to keep us focused on making the most of our time as human beings by doing, saying and being our very best.

Human beings (as homo sapiens) have walked this Earth for only about 200-300 thousand years. That is a tiny fraction of the 4.5 billion years Earth has existed. And of that, the average human lifespan is only around 80 years. That is such a ridiculously short time that we're here, living life in this particular human form. Let's make every moment of this short time, being human, count.

What would it take for you to live it learning all the lessons your life has to teach, with ease and grace? And what would it take for you to help make this experience for all the other human beings around you having their own experience, to be equally joyful and fulfilling?

> *"Beloved child, go out and spread light to the world. Be strong, be brave, be kind"*
>
> In Dreams - Jai-Jagdeesh

15

IT WORKS A TREAT

I've always had a fascination in the various modalities available to people. Once I even naively dreamed of studying everything possible so as to offer clients every option available. I've since learnt the key to treatment and true healing doesn't lie in the modality itself, rather the underlying energy in the subconscious of the individual.

There are countless ways to treat dis-ease. N.B. If you haven't seen disease, broken down as dis-ease, it represents a way of thinking that instead of viewing disease or illness as being something wrong with the body, dis-ease is when the body is no longer in ease and flow. It is a subtle shift in the word but is symbolic of a huge shift in coming from a negative perspective, to viewing the body through a lens of love and acceptance. Dis-ease does not mean something is wrong. There is never anything wrong, as all illnesses and events occur for a greater reason (although sadly these reasons are often beyond our comprehension in our limited human perspective and certainly without the gift of hindsight that time offers). Dis-ease is an indicator that improvements could be made to receive the most out of the experience of being human.

Most treatments act to heal the body. I want to be really clear here that I am not dismissing any form of treatment, as the vast majority work, to some extent, under the right circumstances. Having a healthy and balanced body is a very important layer in this human experience. However, it is important to remember that lessons and growth come from living life in something less than perfect. I am beginning to digress into deeper philosophical discussions about the pursuit of perfection that are best left for other chapters. The intention of this chapter is to present a number of healing modalities, to explain them and unpack why people love and flock to them.

PHARMACEUTICAL BASED MEDICINE - ALLOPATHIC MEDICINE (INCLUDING MEDICATION AND SURGERY)

"Allopathic" can be considered a controversial term by some as it is assumed to be a process of treating symptoms to reverse their presence (sometimes with a medication that induces an opposite response to the symptom itself - i.e. treating constipation with a laxative and vice versa).

Many supporters of complementary medicine oppose allopathic medicine because there is a perception that treatment does not always venture any deeper than the uppermost apparent symptom. This may be true in some instances but it is fair to say that allopathic medicine is primarily focused on the physical body and disregards the presence (or even possibility) of any further energetic layers.

Another huge controversy of allopathic medicine is that it is driven by drug companies i.e. "big pharma" who have the potential for bias towards financial benefits over health. In saying that, there are many regulatory bodies in place to minimise this and we are incredibly lucky to live in an era where there is constantly increasing transparency.

NATUROPATHY

Naturopathy is a health approach based upon the fundamental belief that the body has the ability to heal itself. Each person is viewed as an entire system (as opposed to a finite organ or ailment) and is managed specifically for their own individual needs. Treatment, as the name naturopathy suggests, is based on natural healing utilising both the power of nature and the body's innate ability to heal.

Naturopathy is thought to have originated from 18th century natural healing systems that utilised food, herbs and optimised the balance of the bodies composition and exposure to light, air and water.

Modern naturopathy now encompasses a combination of dietary, nutritional and lifestyle advice as well as herbal medicines, where appropriate, to both treat specific illnesses as well as optimise well-being.

HOMEOPATHY

Homeopathy and naturopathy are often grouped together due to the similarity in their names yet they have distinct differences. Whilst both healing philosophies are based on the belief that the body has the ability to heal itself, homeopathy specifically works on the principle that "like cures like". It is thought that symptoms and illnesses are a natural process that the body undergoes as it attempts to heal itself. Herbs and minerals that would otherwise cause specific symptoms in an individual are highly diluted and then used as a treatment for an illness with similar symptoms. These treatments are designed to trigger the body's natural healing mechanisms.

It is believed the lower the dose (i.e. the more diluted), the more potent the treatments. Homeopathic treatments are often expressed as a dilution factor such as 6X which refers to the potency that is achieved through repetitive dilution and succession according to homeopathic principle. In this case, X refers to a 1 in 10 dilution, 6 times over.

One of the controversies of homeopathy is that some believe that there is so little active substance in the treatment that any perceived benefit is from placebo effect rather than true physical response. Regardless, any potential treatments should always be disclosed to any other health practitioner with whom you are working, regardless of modality.

TRADITIONAL CHINESE MEDICINE

Traditional Chinese Medicine (TCM) is a multifactorial health care system that incorporates Chinese herbal medicine, acupuncture, massage, moxibustion and cupping as well as dietary and lifestyle practices.

TCM originated in China thousands of years ago based on philosophies of health and disease. Modern TCM has been unified as a result of extensive experience and comprehensive clinical observations.

The body is seen to be a holistic system of interconnected energy pathways and organ systems that require delicate balance. This balance is obtained through principles of yin-yang, the five elements and Qi.

TCM is a well established health approach in most of SouthEast China. Although widely practiced and accepted in Western societies, TCM is still viewed by many as an alternate therapy.

There are multiple aspects to TCM which I will explore deeper; however I have listed them each in their own category because, even though they are part of the overall approach of TCM, each can be considered and used as stand alone therapies.

ACUPUNCTURE

Acupuncture involves tiny needles at specific points in the skin around the body. These points correspond to meridians (energy pathways) through which Qi (energy) flows. Acupuncturists believe that inserting needles along these pathways helps to restore the balance of energy flow.

Most commonly, acupuncture is used to treat pain, however it can be used for countless conditions in which energy flow is impacted.

The thought of having multiple needles inserted around the body may sound confronting and quite unpleasant. My experience however is that acupuncture is anything but unpleasant. The needles themselves are so tiny that they are painless and any time I've had acupuncture, I've left feeling absolutely on top of the world (perhaps because my innate balance had been restored). The only time I have ever experienced discomfort was when I was receiving treatment to induce labour. It was explained to me that this was different to other forms of acupuncture. Most treatments are aimed at balancing Qi. However, to induce labour, the opposite effect is needed. Pain is an indicator of unbalanced Qi.

CUPPING

Cupping has seen a rapid increase in popularity recently due to images in the media of celebrities and athletes walking around with telltale circle bruises on and around their backs.

These bruises are caused by suction created by "cups" heated and applied to the skin, causing a vacuum. This vacuum is thought to draw fluid to the surface of the skin to remove toxins and promote healing.

As well as being a key part of TCM, there is also a history of cupping having been popular in ancient Egyptian and Middle Eastern cultures.

The heat applied to the cups before application is quite a pleasant experience however the vacuum created by the cups themselves can become quite uncomfortable. My personal experience with cupping is that the perceived benefits did not justify the huge bruises I was left with. The worst of the bruises took over a week to fade, and ended up being the ones that were visible above the line of my shirt collar. In saying that, I know other people who swear by cupping, especially for muscle pain and tension.

MOXIBUSTION

Have you ever walked past an acupuncturist or traditional Chinese doctor's office and noticed a suspicious smell? Moxa is an herb used in TCM for a therapy called moxibustion. Moxa is usually made from dried Chinese mugwort (*Artemesia aygyi* or *A. vlugaris*) and when it is heated, it smells like marijuana.

The idea of moxibustion is that when a dried stick or clump of moxa is lit and burnt near the surface of the skin, it causes heat to warm and stimulate the flow of Qi in the body. Although it might sound intense and verging on dangerous to light dried plant material near the skin, the dried moxa actually smolders rather than being an active flame.

Moxibustion is a unique aspect of Chinese Medicine as patients can be shown and guided to continue treatment at home between appointments.

AURICULOTHERAPY (ALSO KNOWN AS EAR SEEDS)

Auriculotherapy is a practice I was first introduced to when I was receiving regular acupuncture for fertility and then ongoing pregnancy support. The acupuncturist used tiny little "needles" that resembled seeds with a barely visible little point on them. She placed them around my ear to trigger specific responses (at the time I was being treated for ongoing morning sickness and anxiety). Unlike acupuncture in which the needles are in place only for the duration of the treatment session, I went home after each session with the ear points left in place, covered by a tiny piece of paper tape.

Auriculotherapy is an easy treatment to be performed at home as there are a number of "ear seeds" on the market that do not have needles on them. These seeds work just as effectively by exerting pressure on the relevant points.

Auriculotherapy as a modality, was originally derived from a combination of the principles of acupuncture as well as the work of Paul Nogier, a 1950s French neurologist. It is thought that specific areas around the ear are connected and can stimulate various areas around the body.

REFLEXOLOGY

Reflexology has some overlaps with auriculotherapy as its underlying principle is that the feet, like the ears, have specific points that can be stimulated to trigger certain organs and regions around the body.

Reflexologists apply pressure through a precise form of massage to balance Qi and restore the body to optimum function.

In the 1890s, Sir Henry Head and Sir Charles Sherrington found that there is a neurological connection between areas on the skin and internal organs. Their research has led to the belief that reflexology specifically targets the central nervous system to elicit an overall relaxed state.

Whether it's by balancing Qi or by relaxing the central nervous system, or a combination/overlap of both, reflexology is one my favourite ways to take time out and give myself a little self care.

BOWEN THERAPY

Bowen Therapy originated in Geelong, Victoria by Thomas Ambrose Bowen in the 1950s. He recognised the significant role fascia plays in the body and the impact manipulation of it could have on a patient and their wellbeing.

He found that through a number of gentle manipulations of the soft tissue, he could alleviate the physical complaints of many patients. This evolved to become what is now known as Bowen Therapy. It has now grown to be a modality widely practiced throughout Australia as well as worldwide.

This is a noninvasive technique that involves specific gentle movements over muscles, tendons, ligaments, joints, nerves and fascia. These precise movements stimulate the body's innate healing systems as well as providing relief from musculoskeletal and neurological disorders.

Bowen Therapy techniques have been clinically shown to stimulate proprioceptors (sensory receptors within the body that respond to movement) in muscles which "initiates a brain response, which in turn send nervous system messages back to the fascia to normalise the resting rate of tissues. An essential component of the technique is a 2- to 5-minute pause between sets of moves, which allows for the integration of the messages and the nervous system response. As the tension level is normalised, fluid movements of the lymph and blood are increased in the area, which enhances tissue repair in injury sites" (Hansen and Taylor-Piliae 2011).

KINESIOLOGY

I have resisted the urge to label every modality in this chapter as fascinating, however if one deserves this title it's kinesiology I am so deeply intrigued by this method..

Dr George Goodheart DC, in the 1960's, discovered muscle testing as a reliable way to access information from the body.

His findings combined with ancient principles of Chinese Medicine and Qi, form the foundation of modern kinesiology. The underlying principle is that stressors disrupt the body's innate balance and ability to heal. If a sample of the stress triggering factor is introduced to the patient it will cause measurable biofeedback. That means the body will have a discernible reaction, detected through muscle testing, when a stimulus is introduced to be tested. For example, a kinesiologist may ask a patient to hold a small vial containing a food sample and then look at the way the muscles respond when applying subtle pressures. This process can be done for any stimulus, such as environmental or even emotional. Biofeedback, through muscle testing provides instant information that can then be utilised to implement relevant lifestyle changes.

One of the most beautiful aspects of kinesiology is that it respects and naturally works with the body's innate knowledge, stored in the subconscious, of what it needs to restore balance and facilitate healing. The kinesiologist is merely the conduit to access this wisdom that already exists within each of us.

AYURVEDA

I was first introduced to Ayurveda when I was at university. It was presented to us as an ancient traditional healing method. In the years since, Ayurveda continues to cross my path through my interest in yoga, in food as a healing modality and there has even been a cafe near me that has opened recently based on Ayurvedic philosophies.

Evidence suggests Ayurveda originated in India over 5000 years ago and as such is considered one of the oldest healing sciences. Originally knowledge was passed down directly from master to student but it was only relatively recently that some of this knowledge was transcribed to be accessible for future generations. The Sanskrit translation of Ayurveda is “the science of life”.

The primary emphasis of Ayurveda is to optimise balance in the body in order to optimise health and avoid illness by looking at all factors, both internal and external, that disturb the body's natural balance. These include, but are not limited to emotional and physical stresses, past trauma, seasonal influences and the quality and types of foods eaten, and even how they are cooked and the way in which the body then processes and digests them.

Each person is a unique fingerprint of a combination of three different body types (doshas), vata, pitta or kapha. Quite often, an individual will appear to have a natural dominance of one, or sometimes two body types. Ayurveda aims to restore balance to a person’s natural combination.

Ayurveda is an incredibly interesting and extremely comprehensive approach to health. I don't feel I would be doing it justice in trying to further explain the different doshas and ways in which they can be balanced. That is best left to an experienced Ayurveda practitioner.

Before I move on however, I'd like to share Angela's experience with you. It demonstrates a very simple change that can have profound effects on health. Angela had been experiencing severe gallbladder pain that her medical doctor had advised would need to be surgically removed. Before agreeing to surgery, Angela had a friend tell her about a local Ayurvedic practitioner who had an amazing reputation. After assessing her and looking at the underlying imbalance that had led to her angry gallbladder, he advised her to make a simple change to her diet. Until then, she had been eating lots of salads and raw fruit. He'd proposed that the extra work her body needed to perform to digest the raw foods was putting additional stress on her body and causing imbalance. Instead she was to switch to only cooked vegetables and gently stewing her fruits. That simple change made a huge difference and Angela has not experienced any gallbladder pain since and did not need to have surgery. Angela's experience is her own and Ayurveda is unique to each individual. If you or someone you know is experiencing similar pain, please seek your own advice from someone experienced to treat appropriately. Angela's advice was specific for her and has not been shared to be in place of your own personalised recommendations. It is simply a demonstration of the powerful outcomes that are possible when balance is respected and restored.

COMPASSIONATE INQUIRY

Dr Gabor Maté created the psychotherapeutic technique called Compassionate Inquiry. The easiest way to explain what this means is that it is a little like talking therapy mixed with a somatic exploration of the body, examining the underlying basis for physical manifestations of illness. Dr Gabor Maté's approach evolved from the fundamental belief that all illness stems from stress and trauma. The deep wounds that occur here subconsciously impact and direct our choices and actions/reactions throughout our life. When we go back and explore the root cause of the illness and we reconcile the trauma that occurred, we can heal from illness and release unhealthy patterns and behaviours.

Whilst the thought of uncovering and reliving trauma may sound like a potential minefield of further trauma, Compassionate Inquiry views the instigating events and resultant stress, trauma and subconscious beliefs through a lens of compassion, respect and acceptance. The insight gained through this approach allows the individual a great deal of freedom through the release of long held suppressed emotion. It also allows for a subconscious reset of old patterns and beliefs and the opportunity to freely express ourselves, wholly and completely, often for the first time ever.

POLARITY THERAPY

In researching this chapter I came across Polarity Therapy. Until now I'd never heard of it. From what I've been able to glean, it is a similar approach to Bowen Therapy in that gentle movements of the body that are referred to as rocking, as well as some rotations and stretching to rebalance the body's energy system. In conjunction with movement, this form of therapy also incorporates dietary changes, yoga and self awareness practices.

Polarity Therapy comes under the banner of energy medicine in which the therapist identifies what is causing possible energy blocks and then will place their hands on the positive and negative points in the body to help it "recharge".

I have been told that after a polarity therapy session, patients tend to feel calm, relaxed and have a renewed sense of vitality.

Now whilst I don't know a great deal about this practice, I am very wary of any modality that involves a practitioner manipulating your energy. The human body is the most brilliantly designed machine the world will ever see and it has the capability to rebalance itself. I strongly believe that any healthcare a person seeks should be done so with the intention to nurture and support the body's own innate healing mechanisms.

One of the most important aspects of our human body is for it to communicate our deeper spiritual needs to us. If we go to a practitioner of any form who "heals" us, not only do we give our power away, we also bandaid over the true meaning of the ill health. This goes for any kind of treatment, whether it be allopathic, nutritional or even energy medicine. No practitioner is capable of "fixing us", only we can do that for ourselves. Health practitioners are not our saviours, they are part of our support network.

PSYCHIC SURGERY

I hope it has been conveyed through the pages of this book, that I do my best to be impartial and give you as much credible evidence as possible. I'm about to sidestep from that intention and tell you about one of the most bizarre and intriguing (and possibly ridiculous) treatments that I have come across. Psychic surgery…

A psychic surgeon claims to be able to remove unwanted or unhealthy tissue from a patient with only their bare hands and without causing any visible wound to the patient. Now I like to consider myself very, very open minded, but this practice has me scratching my head.

I have heard mention of it over the years but have never personally come into contact with anyone who has reported having had it performed on them, or even anyone who "knows someone" has had psychic surgery. That in itself is quite abnormal as usually there is always "someone who knows someone", it's the old theory that we're all only six degrees of separation.

Psychic surgery appears to be a practice that was traditionally performed in the Philippines and Brazil. The internet is full of people claiming to have experienced the miraculous healings of psychic surgery but as I said, I haven't been able to get any reliable first, second or even third-hand accounts of it, so it is quite likely a myth. I did come across one recording of a person performing psychic surgery in which there was a very blatant sleight of hand that showed that in this instance, the practitioner (and I use that term very lightly here) was deceitful and betrayed the trust the patient had instilled in them.

Despite my lack of evidence, I still thought it was worth mentioning the practice of psychic surgery here in this chapter as a) absolutely anything is possible as the human body is truly just a projection of our reality and b) the likelihood of this practice being a myth and scam is a great reminder to always be discerning when placing any trust in another along your journey of health.

16

EFT, NLP...WTF?

EFT AND NLP are two acronyms that you may have heard used in passing. Three little letters for two very different approaches that are helping people reprogram the way they see and experience the world to make real and lasting changes in their lives.

EFT

The amygdala is a region of the brain that is very responsive to fear. So much so that we think of it as our fear centre. When a fearful situation comes along, the body releases neural substrates (particularly serotonin, but also dopamine, noradrenaline and acetylcholine) to inform the amygdala that something is going on. The amygdala then triggers the body to release further substrates such as adrenaline and cortisol. Overtime, the amygdala is conditioned to believe that certain situations are always fearful or dangerous and thus release these hormones with very little stimulus. This becomes a default activation system, constantly putting the body under unnecessary stress and possible overwhelm. It is worth noting that people diagnosed with chronic depression have been observed to have an enlarged amygdala, thought to be due to continual excitation and hyperactivation.

Emotional Freedom Technique (EFT), created by Dawson Church in 2010, is also known by the simple term, tapping. Tapping is similar to a form of acupressure in that it activates meridians to send signals to calm the amygdala. This in turn stops the cascade of signals that activate the endocrine system that would normally start producing stress hormones.

By tapping on particular points around the face a body whilst repeating a fear or stress inducing phrase (to induce and replicate a stimulus that invokes heightened negative emotions) EFT is thought to rewire the brain's neural response. When the person is once again exposed to the stressful stimulus, the amygdala has essentially been programmed to no longer recognise the stimulus as stressful. Instead it recognises the sensation of calm and safety that was created during tapping. Safety is especially important for people in distress as it helps create new pathways that do not involve adrenaline and cortisol and the hypothalamus-pituitary adrenal system.

HOW TO TAP

(Please refer to the references at the end of this book for more detailed instructions)

1. Identify a stressful thought, belief or situation in your life right now and make a note of it.

2. Find a comfortable position, either sitting or lying down (however as this is an active process, it is better to be sitting).

3. Take some deep, calming breaths and connect to the emotion/thought/belief that you identified in step 1.

3. Rate the intensity of how it makes you feel on a scale of one to 10. You'll check in afterward to see if the intensity has changed after tapping.

4. Tapping several times on each point, work your way around the tapping points whilst silently or out loud repeating a phrase along the lines of "Even though [insert stressful thought/feeling/belief] I deeply and completely love and accept myself".

5. Repeat the sequence of points around the head and body through three cycles. As you do, allow your emotions to flow freely whilst keeping a calm, steady breath.

6. Repeat a further three rounds of the entire sequence, this time modifying the statement to include the layer of "because" to unpack your thoughts and feelings eg. "Even though [stressful/thought/belief] because [why you feel this way], I deeply and completely love and accept myself".

7. Feel free to adapt the statement as you progress through the points as more and more insight about the feelings you're experiencing may come to light. It is your choice to do as many rounds as you feel necessary as you uncover more and more of what is lying in your subconscious.

8. Choose an affirmation and tap through the points another three times.

9. Once finished, take a few moments to breathe deeply and connect with the sensations and feelings that you're experiencing.

11. Now rate on a scale from one to 10 how you're feeling at the end of the process. The goal here is to notice a significant drop in how you rate the intensity of your feelings. In the rare case that the number increases, this is an opportunity to go back again and explore even deeper what is behind these thoughts/feelings/beliefs.

This is a brief summary of the ways I have been taught to tap by a number of different sources and teachers. Some teachers may guide you through a slightly different process or use different words. Ultimately, there are countless teachers who can guide you and teach you how to tap. This summary is not designed to replace their expertise (and it is for this reason I have not included details of the specific points to tap as I wish to respect the wisdom and experience of the many teachers around the world) so if EFT and tapping is something that you are drawn to, I highly recommend seeking a teacher with whom you resonate.

As with any modality, there are teachers and guides who will simply take you through the process of tapping but there are also a multitude of sources to learn the true depths of EFT and tapping for yourself and even the opportunity to become professionally certified to help others.

NLP

Neuro-linguistic programming (NLP) is a popular modality that started in the 1970s. NLP is designed to help change people's thoughts, beliefs and actions by looking at their perceptions, behaviours and how they communicate (language processing). These are then reviewed, adapted and modified with the intention of easing phobias and anxieties as well as improving overall happiness, well-being and productivity.

Richard Bandler and John Grinder believed it was possible to identify the patterns of thoughts and behaviours of successful individuals and then teach them to others. They published two books, *The Structure of Magic I* and *II*. These books were very well received and launched the awareness of NLP into the mainstream.

The idea behind NLP is that our thoughts act as a form of subconscious programming within us. Throughout our day to day lives, we follow this subconscious programming to create our reality. By consciously using targeted language, NLP works to reset the underlying programming, to reframe the way we perceive and interact with the world around us. NLP has a wide variety of applications and is popular through a wide variety of different communities from highly driven individuals in the corporate world, aiming to be more successful to spiritual seekers looking to instil change in the way they view the world. Rather than being told what to think, NLP teaches people how to think to get the most out of life.

Given the relative infancy of NLP, and the fact that it's based on theoretical science rather than empirical science it is hard to find completely unbiased evidence to support its foundation. Yet every individual I have spoken with who has experienced NLP, raves about it and its benefits.

My own brief encounter with NLP blew my mind. I grew up in a rural area on Queensland's Sunshine Coast.

surrounded by venomous king brown snakes. And when I say surrounded, I'm not exaggerating, not a summer would pass without having a close encounter of nearly stepping on one walking just a few steps from the house. As a result, I became terrified of snakes. To me, they meant danger and possible death. I learnt to look everywhere for them and constantly feared they were hiding around every corner. As an adult living in the city, my fear of snakes became something I could just ignore. The only time it would ever resurface was when I went home to visit mum and dad.

A few years ago I went on a retreat that happened to be held about 30 minutes from my parents home, back in brown snake territory. And unlike my parents' property that is relatively clear of bushland and places for snakes to hide, this retreat was actually in the bush itself. My fear of snakes walking to and from the retreat hall to my room everyday had me so terrified that I struggled to get the most out of the retreat. A few days in, the retreat facilitator (also an NLP practitioner) recognised my fear and talked to me about them. Truly I don't remember what he actually said but I do remember answering lots of questions of the thoughts and feelings that come to mind about snakes and the next thing I knew, the snake phobia was gone. Since then I have held a snake (not a venomous one), I've been able to look at a snake and appreciate its beauty and I even have a sacred piece of artwork in my office, centred around a snake. My own brief experience was enough to convince me that there is some powerful potential in NLP.

Again, if this is something that deeply interests you, I would highly recommend seeking a guide or teacher qualified to take you through the process.

17

BEYOND THE SUPERSTITIONS

Ricci-Jane Adams, of the Institute of Intuitive Intelligence, describes many of the things that are believed to have power in the world of spirituality and mysticism (such as crystals and tarot cards) as the "Trinkets and superstitions of spirituality".

One of the key presumptions of spirituality and mysticism over the years, is the belief that inanimate objects hold power. Crystals, beads, incense, statues, tarot and other types of cards, rituals and astrology, hold no power. All power comes from within and from our connection to consciousness. We are the power. Everything we will ever need is already within us.

I'm sure there are people who will swear that they have had a tarot reading and it predicted mind blowing things, or perhaps their astrological reading is always incredibly accurate, or maybe even since wearing a certain crystal in their pocket that they've been happier/luckier/healthier... All of this is totally and completely valid, I'm not here to dispute that. it's not the object or the act in itself that has created an outcome, it's the intention of the individual. The emotions we feel when holding a specific object or the feelings and intentions of a tarot card reader, direct the outcome. This is where the power lies, not in the object, but in the individual. We are our own source of power. There is no separation between us and the energy around us, we don't need something outside of us to impact that energy.

As covered in earlier chapters, our thoughts become our reality. If we hold the vision of a certain outcome, under the right circumstances, we create the outcome. We manifest it. This goes for both positive and negative intentions. If we focus our attention on fearing something, then we direct our energy towards manifesting the feared outcome. If we step out of fear and into the energy of love, focusing our attention and intentions towards outcomes that fill our heart with love and joy, then we manifest our desired outcome.

Manifest, and the concept of manifesting is a word thrown around a lot in new age circles. I've spent considerable time myself trying to master the art of manifestation.

Dr Joe Dispenza (a groundbreaking neuroscientist in the field of neuroplasticity and the ability to change your life through your thoughts) has a number of guided meditations. In these, he takes you on a journey of rehearsing the emotions you will experience if your desired goal comes to fruition. His reasoning is that our energy doesn't know the difference between something actually happening and us simply cultivating all the feelings as if it had. A lot of people are familiar with mentally rehearsing, but this is more appropriately called emotional rehearsing. Dr Joe's catchphrase is "the feeling is the prayer". It took me a little while to understand the depth of this statement. Basically it means that we don't simply wish for something to happen. Instead we drop down into our true powerhouse, our heart, and foster the feelings as if the outcome has already occurred. Dr Joe's saying reminds us that feeling the emotions directs the outcome. The emotive heart-brain is in direct communication to our subtle energy field, surrounding our body as well as the universal energy field beyond. Our heart and the feelings we hold in it, influence everything around us.

It is my hope and wish that after reading this book you will have a greater appreciation for yourself and the gift of your existence. Being human is the reason we are here. As you will have read, a great number of brilliant minds and hearts have contributed to helping us gain a greater understanding of how and why we exist in the world. We are no longer bound by the fears and superstitions of our predecessors. We have the opportunity to make the most of this human experience and gain the most of our time on Earth.

With this in mind, I urge you to use your discernment wisely if choosing a guide, mentor or teacher to work with in this space. Please check in with your heart and your own beliefs before making any decisions.

There are a great many people who are at a low point in their life who look to spiritual teachings and new age to help find a way out of their situation. This is absolutely the right direction to look, but sadly this also means that there are some opportunistic operators who prey on people when they're most vulnerable. Equally, there are many people who have found a lot of what they deem to be answers whilst working through their own shadows and demons, and are eager to share this wisdom with the world. No-one is perfect and there are always risks that their shadows are still dominant in their life and clouding their perceptions. I don't say this in judgement or to scare you off working with guides/mentors, only to remind you to check in with yourself, use your intuition and see what feels right. What is ideal for someone else may not be the best match for you. When two people work together, the combination of energy creates a unique alchemy. Always ask yourself if the alchemy feels right.

When a guide, mentor or teacher works with someone, they receive just as much wisdom and enlightenment from the client, as they give. This interplay between the two is the very crux of alchemy. The perfect energetic combination of two individuals creates a very specific and unique outcome. This is another reason to closely follow your intuition in these instances, it will always guide you to the right person.

I don't propose to have all the answers, I have tried to present the information as I've come to understand it, as best I can (and as straightforward as possible). We are constantly evolving, learning new things and understanding our human existence deeper and deeper. I will, however, warn against anyone who claims to have all the answers. I believe that as spiritual beings, having a human experience, we will never have all the answers. We will only find out the full extent of the truth once we graduate this Earth School, and wholly return to the Universal Consciousness.

18

CONSPIRACY THEORIES

If you spend any time around groups of people seeking spiritual truths, you're sure to find some people who fit the stereotype of being tin-foil hat wearers. Perhaps that is a little extreme but I'm sure you know the people I mean, always looking for some alternative agenda within the government, trying to find some sort or ulterior motive behind anything in the media or popular in society. People who are constantly engaged in the drama of conspiracy theories.

Conspiracy theories and the propagation of misinformation were a huge motivating factor for writing this book.

Information is now more readily available than ever before.. With the freedom of accessing this knowledge also comes the equal freedom to share it with others. Thanks to the advancements of technology, anyone and everyone has the opportunity to present themselves as an apparent expert. Cue the circulation of conspiracy theories.

The definition of conspiracy theory is an explanation that suggests a negatively skewed cover-up of the truth, often against mainstream and more probable beliefs.

I guess you could call the world we're living in right now as a world with a spectrum of beliefs. At one end of the spectrum, people who only believe what can be validated with traditional empirical science. It is these people who are most likely to be dismissive of some of the concepts I've attempted to explain here. One the very opposite end of the spectrum there are people who approach all information with a level of suspicion. There is also a tendency to get caught up in, and very emotionally invested in their suspicions that in extreme cases verges on paranoia.

Historically we've seen huge theories emerge about certain powerful governments covering up evidence to prove the existence of alien life forms or suspicions about political assassinations.

More recently it's been the rollout of 5G (the fifth generation of mobile network), and now speculation about the origin of Covid-19 and vaccinations against the virus. There may be elements of truth to the information circulating, there are also some extremely concerning theories being shared. One in particular has suggested that anyone who is vaccinated will now be microchipped and biologically coded to link to cryptocurrency in the future. Whilst we can't conclusively prove or disprove theories such as these at the moment, I think it's fair to say that it sounds completely ridiculous and not worth investing any further unnecessary energy.

Whether or not any of these theories are true is beside the point. Grasping onto the belief that the Government or other powerful organisations are purely motivated by controlling the masses and suppressing the truth, is dangerous. Whilst it's good to have a constantly questioning and seeking mind, buying into fear and speculation will do you more harm than good. It is not the theories that are the problem, purely the energy of fear and suspicion that surround them.

I'm not here to pass judgment on any idea, well perhaps I am on the cryptocurrency one, but all others are simply just theories until they are proved otherwise.

I believe, and it is one of my strongest values, that we all have the freedom to choose. The key is to make your choices from an empowered and heart centred place.

Where attention goes, energy flows. Buying into the beliefs and fears of conspiracy theories channels your energy and awareness to that space. You create your reality. If you believe in conspiracy theories then that is the reality you will create. Personally I set the intention that governments and high powered organisations in the world are composed of human beings such as myself, doing their best with the tools they have. Admittedly, money and power make people do unexpected things, but at the end of the day, we are all having an experience of being human, connected by invisible threads of energy that are Divine Consciousness.

So I implore you to always keep an open, questioning mind, follow your own internal guidance and always check in with your heart-brain to see what information feels right for you.

To quote two very famous pop culture references of my time.

"The truth is out there."

Mulder and Scully - The X-Files

19

REFERENCES

THE SCIENCE CHAPTER

HTTPS://WWW.BRITANNICA.COM/SCIENCE/ATOM

https://chem.libretexts.org/Bookshelves/Physical_and_Theoretical_Chemistry_Textbook_Maps/Supplemental_Modules_(Physical_and_Theoretical_Chemistry)/Atomic_Theory/The_Atom/Sub-Atomic_Particles

https://theconversation.com/can-consciousness-be-explained-by-quantum-physics-my-research-takes-us-a-step-closer-to-finding-out-164582

https://plus.maths.org/content/physics-minute-double-slit-experiment-0

https://www.amazon.com.au/Clair-What-Understanding-Clairsentience-Claircognizance/dp/1475124147

IT'S THE LAW

https://www.mind-your-reality.com/seven_universal_laws.html

SPIRITUAL BEINGS HAVING A HUMAN EXPERIENCE

https://www.verywellmind.com/what-is-the-ego-2795167

https://www.youtube.com/watch?v=klpDHn8viX8

https://www.nderf.org/Experiences/1anita_m_nde.html

https://www.anitamoorjani.com/

THE HEART KNOWS

https://pubmed.ncbi.nlm.nih.gov/31739081/

https://www.heartmath.org/research/science-of-the-heart/

https://www.heartmath.com/science/

I'VE BEEN HERE BEFORE

https://www.jimbtucker.com

https://www.healthline.com/health/mental-health/what-causes-deja-vu

https://www.newscientist.com/article/2101089-mystery-of-deja-vu-explained-its-how-we-check-our-memories/

https://pubmed.ncbi.nlm.nih.gov/15654548/

THE INFINITE WISDOM OF THE BODY

-

WHEN ILLNESS STRIKES

-

GRADUATION DAY

-

IT'S A VIBE

https://www.researchgate.net/publication/342562977_A_Brief_History_of_the_Chakras_in_Human_Body

https://www.amazon.com/Anatomy-Spirit-Seven-Stages-Healing/dp/0609800140/

https://www.nasa.gov/mission_pages/sunearth/news/gallery/schumann-resonance.html

HO'OPONOPONO

https://www.hooponoponoaustralia.com.au/about.html

https://www.huna.org/html/ponorevealed.html

MEDITATION

https://www.verywellmind.com/mindfulness-meditation-88369

https://mindworks.org/blog/what-is-spiritual-meditation/

https://www.verywellmind.com/practice-focused-meditation-3144785

https://aihcp.net/2012/10/17/what-is-movement-meditation/

https://www.healthline.com/health/mantra-meditation#benefits

https://merakalpamalas.com/pages/the-anatomy-of-a-mala-necklace

https://www.tm.org/blog/meditation/happy-new-year-how-the-transcendental-meditation-technique-can-make-it-even-better/

Transcendental Meditation, Self-actualization, and Psychological Health: A conceptual overview and statistical meta-analysis Alexander, C. N., Rainforth, M. V., & Gelderloos, P., *Journal of Social Behavior & Personality*, 6(5), 189-248, 1991.

https://ggia.berkeley.edu/practice/loving_kindness_meditation

https://observer.com/2017/06/neuroscience-mindfulness-brain-when-you-meditate-development/

https://www.yogitimes.com/article/what-are-mudras-healing-fingers-yoga

LEARNING TO GO WITHIN

https://www.heartmath.org/articles-of-the-heart/spiritual-heart/

https://www.spotify.com/

RELIGION

https://en.wikipedia.org/wiki/Religion

https://en.wikipedia.org/wiki/Christianity

https://en.wikipedia.org/wiki/Islam

https://en.wikipedia.org/wiki/Hinduism

https://en.wikipedia.org/wiki/Buddhism

https://en.wikipedia.org/wiki/Bahá'%C3%AD_Faith

https://iep.utm.edu/gnostic/

THE BALANCING ACT

https://www.miguelruiz.com/the-four-agreements

https://www.thefouragreements.com/how-to-use-the-four-agreements-to-create-a-better-life/

IT WORKS A TREAT

https://www.healthline.com/health/allopathic-medicine#controversial-term

https://www.australiannaturaltherapistsassociation.com.au/Public/ANTA_Therapies/Naturopathy

https://www.webmd.com/balance/what-is-homeopathy

https://www.sitcm.edu.au

https://www.mayoclinic.org/tests-procedures/acupuncture/about/pac-20392763

https://www.health.harvard.edu/blog/what-exactly-is-cupping-2016093010402

https://www.takingcharge.csh.umn.edu/explore-healing-practices/moxibustion

https://dantianhealth.com.au/what-is-moxibustion/

https://www.sciencedirect.com/topics/pharmacology-toxicology-and-pharmaceutical-science/auriculotherapy

https://www.takingcharge.csh.umn.edu/explore-healing-practices/reflexology/how-does-reflexology-work

https://www1.health.gov.au/internet/main/publishing.nsf/Content/phi-natural-therapies-submissions-containerpage/%24file/Bowen%20Association%20of%20Australia.pdf

https://pubmed.ncbi.nlm.nih.gov/22087611/

https://www.aka.asn.au/about-kinesiology

https://www.ayurveda.com/resources/articles/ayurveda-a-brief-introduction-and-guide

https://compassionateinquiry.com/online-training

https://www.naturaltherapypages.com.au/energetic_medicine/polarity_therapy

https://doh.gov.ph/sites/default/files/news_clips/050719-0004.pdf

EFT, NLP… WTF?

https://www.thescienceofpsychotherapy.com/glossary/amygdala/

https://kripalu.org/resources/how-use-tapping-calm-anxiety

https://www.eftuniverse.com/

https://www.goodtherapy.org/learn-about-therapy/types/neuro-linguistic-programming

BEYOND THE SUPERSTITIONS

https://www.riccijaneadams.com/

https://drjoedispenza.com/

CONSPIRACY THEORIES

-

ACKNOWLEDGMENTS

All the work presented here is an accumulation of multiple brilliant teachers. Unless stated otherwise, none of the contents of this book are my original discoveries. My intention has been to present all the information as I understand it, in a succinct and graspable manner. I have included references and links to further reading where relevant.

Please also know that if I have inadvertently missed acknowledging a source, it has been unintentional. Some information that I have been taught along my life journey, has felt so familiar, as though it has awoken a knowing in my soul that was always there, I don't always recall where I was first exposed to the concept. This is part of my claircognizance.

I have had many teachers through this life thus far. Most in the form of friends and family but unexpectedly one was a deeply religious teacher who taught me so much more than high school biology. He showed me that science and faith can coexist.

A huge thank you goes to Aunty Sheila (an honorary aunt because of the special place she holds in our family). You hold the wisdom of the universe in your tiny frame yet role model ultimate humility in never flaunting it or believing you have all the answers. I am also so deeply grateful to have you as part of my cheer squad and editorial team for this book. I wouldn't have had the courage to write Spiritual Being Human without you in my corner. I am also so appreciative of her late husband, my Uncle Doug who taught me from a young age that there is so much more to health and wellness than the eye can see. And that you're never too old to keep learning

Thank you to Ricci-Jane Adams for being a fearless teacher and putting herself so boldly out into the world for me to find when I needed a teacher.

To my dear friend Barbara, I am so grateful you saw my internal light when it was lost in the haze of the early days of parenting, and for giving me gentle (and not so gentle) nudges when I've needed them, for giving me permission to step into my truth and for being my sounding board ever since.

Thank you to my Mum for being my biggest champion through life. Thank you for your unconditional love, for always believing I can achieve anything and for encouraging me to always reach for the stars and beyond.

My Dad Kerry, I can only imagine how difficult it is to be a stepparent. Although I was generally a good kid, I know that some days I didn't make it easy. Thank you for coming into our lives, for falling in love not just with my mum, but also with me. Thank you for stepping up when I needed a father and for graciously stepping back just as often so I could still have a wonderful relationship with my biological father. Many people feel broken from being divorced children, I have never felt anything but blessed and it is thanks to the gift of having you in my life. Thank you for opening my eyes to so much of spirituality from such an early age.

To my Dad Paul, thank you for encouraging and nurturing my scientific and enquiring mind. For all the trips to the powerhouse museum on holidays with you, for the gifts of science kits and all the wonderful books opening my mind to the world of words. For nurturing and encouraging reading and my love of books. Now that I have children myself, I can appreciate the significance of giving me The Hobbit and then The Lord of The Rings trilogy when I was ten.

I can't write a list of special people in my life and the role they've played without mentioning my sister and grandmother. You are both immensely powerful women and have shown me the beauty and strength in divine feminine energy.

To my husband Will for allowing me the time and the space to explore my own spirituality and to write this book, thank you for pushing me when I've needed it and for constantly encouraging me to grow.

Above all, my two greatest teachers are my children. Thank you for choosing me as your mum. To my son, who pushes me to grow every single day. constantly challenging me, who one minute takes me to the ends of my coping skills and then the next, absolutely floors me and leaves me in awe of his own wisdom and brilliance. Thank you for showing me the fragility in your brilliance, for being the example of a new, wise generation coming through. Thank you for reflecting my own depth and seriousness.

To my daughter, thank you for showing me what it is to be free and live in absolute joy. I am constantly in awe of your spirit, creativity and determination to experience life your way. Thank you for showing me what it truly is to live from your heart.

A very special final thanks is reserved for the lady without whom this book wouldn't be possible. Spiritual Being Human has been an idea and a calling for many years but I never had the courage to make a reality. When my dear friend Brigid announced that she was starting the Art of Grace Publishing House, I knew at that moment that my book had found a safe and loving home and would be ready to start coming through. Thank you.

ABOUT THE AUTHOR

Since childhood, Kelly dreamed of studying every modality possible to learn and understand health and healing from all perspectives, and then find its most profound truth.

Her dreams led her to study a Bachelor of Pharmacy, believing a pharmacist to be the most easily accessible health practitioner in the community, and be best placed to share her broader health perspective.

Her diligence at university and then focus on establishing a career, as well as going on to own her own business, saw Kelly's dream of becoming a truly holistic expert, encompassing all modalities of health, fall away in the distance.

After having her first child and struggling to hold her perfectly constructed little world together, Kelly was forced inward to cope. Here she connected to her deepest truth, her intuition.

Since embracing her intuition as her guiding light, Kelly has come home to her original goal of viewing health and healing from a deeper perspective, yet the need to study multiple modalities is no longer necessary as she now sees the truth is within us all.

Kelly now supports clients along their illness/wellness journey. As an intuitive health guide, using her unique combination of medical intuition and health transition doula, she guides clients to see that our bodies are always working in our favour. The pain and suffering we experience occurs when we reject or ignore the messages within.

Living on Sydney's Northern Beaches with her husband and two children, Kelly can often be found first thing in the morning sitting at her favourite headland, watching the waves roll in.

www.kellypaardekooper.com
@kellypaardekooper - Facebook & Instagram.

Made in United States
Troutdale, OR
11/03/2026

56845399R00096